A FOUNDATION
FOR QUANTUM
CHEMISTRY

A FOUNDATION FOR QUANTUM CHEMISTRY

A. R. Denaro, M.Sc., Ph.D., F.R.I.C.

A HALSTED PRESS BOOK

JOHN WILEY & SONS
New York–Toronto

English edition first published in 1975 by
Butterworth & Co (Publishers) Ltd
88 Kingsway, London WC2B 6AB

Published in the U.S.A. and Canada by
Halsted Press, a Division of John Wiley & Sons Inc.,
New York

© Butterworth & Co (Publishers) Ltd., 1975

Library of Congress Cataloging in Publication Data

Denaro, A   R
   Quantum mechanics

"A Halsted Press Book."
Bibliography: p.
1. Quantum theory.   I. Title.
QC174.12.D45   1975   530.1'2   75-1286
ISBN 0-470-20880-5

Printed in Great Britain

PREFACE

Most chemistry students are given a qualitative or semiquantitative description of the results of quantum mechanics at an early stage in their courses as an introduction to atomic and molecular structure. Accounts of quantum chemistry which students may subsequently encounter necessarily contain a degree of mathematical formalism. In order to cover a reasonable width and depth in the subject within the confines of a book, the approach of some accounts is such that it tends to dismay the student who does not have a strong background of physics and mathematics.

This book is aimed at those students whose appreciation of physics and mathematics is perhaps not as great as is required for reading quantum chemistry as presented in specialist monographs. It is intended to bridge the gap between the elementary qualitative or semiquantitative descriptions of quantum chemistry and those more specialist treatments found in books dealing specifically with atomic and molecular structure, valence theory and spectroscopy.

It is assumed that the reader has a qualitative knowledge of atomic and molecular structure together with the concepts of elementary differential and integral calculus. Some space is devoted to the physics of waves which is necessary to develop an appreciation of the properties of wave equations. Mathematical operations are treated fairly fully for the benefit of those readers who may lack a facility in such operations. To assist further, the trigonometric relationships and the differential and integral forms which are required in the book are provided in appendices from which the memory may be refreshed. Another appendix provides the necessary but limited amount of information on complex numbers.

The detailed approach used in this book limits the amount of material which can be covered and only problems for which there are exact quantum mechanical solutions are considered. No account of the approximate methods of quantum mechanics is given as this book does not aim to provide a comprehensive grounding in quantum mechanics. Its objective is to give the reader an insight into quantum mechanics and to put him in a position to tackle the more specialist accounts without being daunted by mathematical formalism. As the title implies, the book is intended to provide a foundation for the subsequent study of quantum chemistry.

My thanks are due to several of my colleagues for helpful discussion and in particular to Dr. A. T. Vincent who also wrote the computer programmes for the off-line incremental plotter which provided the plots for *Figures 2.1, 2.2, 2.3, 2.4, 3.2, 3.3, 6.2 and 7.2.*

A. R. D.

CONTENTS

1 INTRODUCTION 1

2 WAVE EQUATIONS 8

Simple harmonic motion and wave equations. Relation between energy and amplitude. General differential equation of simple harmonic motion. Linear combinations of solutions. Alternative solutions of the differential equation. Progressive waves. Standing waves. The Schrödinger wave equation. Eigenfunctions and eigenvalues. Interpretation of ψ

3 PARTICLE IN A BOX 28

Particle in a one-dimensional box. Normalisation of wave functions. Orthogonality. Quantum mechanical operators. The postulates of quantum mechanics. Particle in a rectangular three-dimensional box

4 POTENTIAL ENERGY BARRIERS 54

Directional implications of wave equations. Single potential barriers. The tunnel effect

5 APPLICATIONS OF ONE-DIMENSIONAL MODELS 71

Conjugated systems. The vibrational energy of a diatomic molecule. The electronic energy of a diatomic molecule

6	THE LINEAR HARMONIC OSCILLATOR	91

Approximate solution. Exact solution

7	PARTICLE ON A RING	101

8	THE RIGID ROTATOR	111

The F equation. The T equation. The energy levels

9	THE HYDROGEN ATOM	122

The F equation. The T equation. The R equation. The energy levels. The angular momentum of the electron. The hydrogen wave functions. The radial function. The angular function. Electron spin

APPENDIX 1	Complex Numbers	145
APPENDIX 2	The Tunnel Effect	148
APPENDIX 3	Trigonometric Relationships	151
APPENDIX 4	Differentials and Integrals	153

FURTHER READING	155
INDEX	157

1

INTRODUCTION

The problem facing scientists at the end of the nineteenth century was to postulate a structure for the atom which would account for the nature of atomic and molecular spectra and which would give some understanding of the way in which atoms bond together to form molecules. The first clue to atomic structure was provided by J. J. Thomson (1897) when he showed that the negatively charged electron was a constituent of all matter. As a result of this observation, he postulated, in 1904, a model of the atom which consisted of a sphere of positive electricity with electrons embedded in it, the whole atom being electrically neutral. About the same time (1903) Lennard suggested that most of an atom consisted of empty space and that the material part consisted of neutral doublets, presumably made up of an electron and a positively charged particle. There were many difficulties associated with these models and they are of interest only as the starting point of the development of modern theories of atomic structure.

The most fruitful idea of atomic structure stemmed from the experiments of Geiger and Marsden (1909) on the scattering of α-particles by thin films of metal. From an analysis of these results, it was Rutherford in 1911 who showed that the positive charge of an atom was concentrated on a minute and massive nucleus. Rutherford thus proposed the planetary model of the atom in which the tiny nucleus was surrounded at a relatively large distance by the appropriate number of electrons to give a neutral atom. To account for the fact that the electrons did not fall into the nucleus as a result of electrostatic attraction, Rutherford had to propose that the electrons rotated rapidly about the nucleus.

It was pointed out, however, that such a system would not be stable. An electron moving in a circular orbit with constant angular velocity undergoes a continuous acceleration towards the centre of the circle and classical electromagnetic theory requires that the energy of the electromagnetic field associated with an accelerating charged particle should change continuously. According to the classical theory then, the electron should continuously radiate energy and the radius of its path should decrease continuously until it spirals into the nucleus.

This difficulty, resulting from the application of the ideas of classical physics was symptomatic of many that these ideas were encountering at the beginning of the twentieth century. Classical theoretical physics could be said to begin with Newton who was the first to work out the laws of mechanics and his law of gravitation is still valid on a macroscopic terrestrial scale. The principles of mechanics were applied to optics (amongst other things) in terms of Newton's corpuscular theory of light. The mechanical theory of light subsequently lost favour and was replaced by the wave theory of light at the beginning of the nineteenth century. In 1864, Maxwell put forward his electromagnetic theory and showed that light was electromagnetic radiation. Many physical phenomena could be given explanations in terms of Newton's mechanics or Maxwell's electromagnetic wave theory in which the energy of any system was continuously variable and could have any value. It has been pointed out above, however, that the application of these classical concepts leads to the defeat of Rutherford's planetary model of the atom.

There were originally two other difficulties for classical physics at the beginning of the present century. One was concerned with optics, the Michelson–Morley experiment, which is not particularly relevant to the topic of this book. Suffice it to say that the difficulty was overcome by Einstein's proposal of the principle of relativity which has shown Newton's law of gravitation to be a very close approximation of a more general case. The other significant problem was that associated with the energy of the electromagnetic radiation emitted by a perfect 'black body' radiator. The energy of the radiation emitted is not uniform but has a maximum value at a definite wavelength which is inversely proportional to the thermodynamic temperature. Attempts to account for this behaviour on the basis of classical physics by Wien (1896) and Rayleigh (1900) were not entirely successful. Rayleigh's treatment was based on the assumption that the radiation was emitted by oscillators and it was an unavoidable conclusion from classical theory that the oscillators could have a continuous range of energies.

It was Planck in 1900 who resolved the difficulty by making the entirely new suggestion that the energy associated with the oscillators is not continuously variable but can consist only of an integral number of quanta, the energy, E, of a quantum depending on the frequency, v, of the radiation such that

$$E = hv \qquad (1.1)$$

where h is Planck's constant. Moreover, Planck suggested that energy could only be emitted or absorbed in quanta. On the basis of this quantum theory, Planck was able to explain completely the behaviour of black body radiators.

In 1905 Einstein was able to lend support to Planck's quantum theory by applying it to the photoelectric effect in which electrons are released from a metal surface *in vacuo* when it is illuminated by light of certain frequencies. The nature of the photoelectric effect was completely at variance with the classical wave theory of light but Einstein showed that the behaviour could be explained by regarding light of frequency v as consisting of a stream of photons of a particulate nature where each photon had a quantum energy of hv. On this basis, Einstein deduced an equation relating the kinetic energy of the ejected electrons to the frequency of the incident radiation and this equation was subsequently verified experimentally by Millikan in 1916.

Further evidence for the photon nature of radiation was provided by the experiments of Compton (1923) in which X-rays were scattered by various materials. It was found that the scattered X-rays contained, in addition to those having the incident wavelength, others of a somewhat longer wavelength. Since the scattering is produced by the electrons in the target material it is these electrons which are responsible for the increase in wavelength of the radiation. Once again this effect may be explained by considering that the radiation consists of photons of quantum energy hv. As a result of the interaction, the electron recoils and the wavelength of the scattered radiation may be calculated from a consideration of the energy and momentum balance for the process.

The success of the quantum theory and the applicability of equation 1.1 leads to the reflection that the sharply defined lines of atomic spectra must be associated with definite energies. With this in mind, Bohr adopted Planck's ideas and applied them to the hydrogen atom. In 1913 Bohr postulated that the motion of the electron in the hydrogen atom was restricted to a number of circular orbits with the nucleus at the centre. An electron moving in one of these orbits had a constant energy and the size of the orbits was determined by the *arbitrary* assumption that the angular momentum of the electron about the nucleus was quantised in integral multiples of $h/2\pi$. The angular momentum could thus be $nh/2\pi$ where n was an integer and was called the quantum number. Radiation was considered to be emitted when an electron moved from an orbit of higher energy, E_2, to an orbit of lower energy, E_1, when the frequency, v, of the radiation was given by

$$v = (E_2 - E_1)/h \qquad (1.2)$$

On the basis of this model, Bohr could explain the observed spectra of hydrogen atoms with a large measure of success.

With spectrometers of high resolution, however, it was found that there were more lines in the hydrogen spectrum than could be explained by the Bohr theory. To account for this discrepancy, Sommerfeld in 1915 postulated that the electron could move in ellipses of various eccentricities. To define the quantum restrictions on the angular momentum of an electron moving in an elliptical orbit, two quantum numbers are required so that Sommerfeld postulated an additional arbitrary quantum number.

A further difficulty was that in the presence of a strong magnetic field, additional lines appeared in the spectrum. This effect was first observed by Zeeman (1896) before the advent of the quantum theory. It was considered that the plane containing the electron orbit could only take up certain orientations with respect to the magnetic field and to specify these quantised orientations a third arbitrary quantum number was required.

With these arbitrary additions of extra quantum numbers, the Bohr theory was very successful in providing an account of the spectra of one-electron species such as the hydrogen atom and the helium ion but it failed completely in the case of the helium atom and other multi-electron systems. There are also other objections. The path of the electron was determined by the methods of classical physics which were then ignored in postulating the quantisation of momentum. The values of the quantum number postulated by Sommerfeld turned out to be incorrect and there was no way of calculating the binding energy of even a simple molecule like the hydrogen molecule. Again, the use of planar orbits would imply, at least for the hydrogen atom, a flat atom which is contrary to the results of experiment.

The Bohr theory was important in that it introduced the idea of quantum restrictions to atomic structure but a more acceptable theory would have to be based on fundamental hypotheses which would lead in a logical manner to the quantisation of energy, angular momentum and other properties.

Two approaches have been made to the problem. One of these is the matrix mechanics of Heisenberg (1925) which involves no atomic model. In observing the behaviour of electrons or atoms, the position or velocity of an electron in an orbit is not directly observed. It is in the equations connecting these unobserved quantities that

INTRODUCTION 5

difficulties arise. Heisenberg omitted these details from his equations and developed a system of calculation in which only observed quantities occur. The method is analogous to that of classical thermodynamics in which the behaviour of systems can be predicted from thermodynamic equations which require no knowledge of molecular structure. Guided by the same principles Dirac (1928) also applied matrix analysis to the problem.

The other approach is the wave mechanics of Schrödinger (1926) and since the ideas underlying this treatment have become important in chemistry it is the purpose of this book to provide an appreciation of the method.

The key was originally provided by de Broglie. Einstein's work on the photoelectric effect had shown that electromagnetic radiation could be regarded as corpuscular, the particles being called photons, the energy of the photon being given by equation 1.1. The radiation also must have the character of waves as it is only on the wave theory that the diffraction and interference phenomena can be explained. Now the product of the wavelength, λ, and the frequency, ν, of any wave motion is equal to the velocity of the wave and hence for electromagnetic radiation

$$\nu\lambda = c \qquad (1.3)$$

where c is the velocity of light. Further, Einstein's theory of relativity provides an equivalence between energy E, and mass, m, $E = mc^2$, so that the momentum, mc, of the photon is given by $mc = E/c$. Writing the momentum as p and substituting for E and c from equations 1.1 and 1.3 respectively,

$$p = h/\lambda \qquad (1.4)$$

so that a relationship exists between the momentum of a photon and the wavelength of the radiation. Equation 1.4 has come to be known as the de Broglie equation.

The properties of radiation could thus, in some respects, be described in terms of the momentum of the particles. It was de Broglie in 1924 who suggested that equation 1.4 might also apply to particles such as electrons and that their behaviour in some respects might be described by equations that are normally associated with waves.

At a later date this idea of de Broglie was verified experimentally for low energy electrons by Davisson and Germer (1927) who obtained diffraction patterns for electrons reflected from a crystal. Subsequently, G. P. Thomson and Reid (1928) also obtained diffraction patterns from high energy electrons which had penetrated thin films of metals. The patterns obtained on a fluorescent screen are such that there are bright concentric circles of maximum intensity, the intensity of fluorescence on other parts of the screen being much lower. The concept also applies to atoms as Stern (1930) and Rupp (1932) have shown that beams of hydrogen and helium can produce diffraction patterns.

Through the impetus of de Broglie's suggestion, Schrödinger (1926) proposed a new description of the atom in terms of waves rather than the material particles of Bohr's model. The situation was that spectroscopists had shown that distinct lines characterised atomic spectra. Bohr and others had emphasised that any adequate model of the atom must incorporate the quantum ideas of Planck and that this requirement could be met by arbitrarily introducing quantum numbers.

Schrödinger, having the clue of de Broglie's suggestions about the wave nature of electrons, realised that the standing waves associated with vibrating strings and other objects were mathematically expressible by equations in which a set of integers necessarily and naturally appeared. He needed only to find the right vibrating object as a model and the standing waves for this object would have equations containing a set of integers. The equations could then represent an electron in an atom and the integers would be the quantum numbers.

The mathematical work of Hamilton provided the right model. Hamilton was working on a problem concerned with tides and had idealised the problem by writing the mathematics of the vibrations of a uniformly deep ocean over the surface of the earth. Schrödinger realised that the solutions of the equations for this model were the equations of standing waves and that these equations could represent the electrons in an atom. The Schrödinger treatment thus has to regard the atom as a purely mathematical concept without requiring any definite model as a basis for the calculations. It does, however, generate quite naturally the three quantum numbers mentioned previously in this chapter.

There are, of course, four quantum numbers required to specify the state of an electron. The necessity for the fourth quantum number arose from the fact that many spectral lines were actually multiplets consisting of two or more lines close together. It was Goudsmit and Uhlenbeck in 1925 who suggested that the electron has an intrinsic angular momentum or spin which contributes to the total angular momentum. This contribution was supposed to be quantised thus requiring a fourth quantum number for its specification. Although this quantum number does not occur naturally in Schrödinger's treatment it can be dealt with in a manner analogous to that used for the magnetic quantum number which does come from Schrödinger's wave mechanics and there is no difficulty.

INTRODUCTION 7

It may be mentioned that Dirac's quantum mechanics does generate naturally all four quantum numbers and has shown that Schrödinger's treatment is a special case in much the same way as Newtonian mechanics is a special case of Einstein's relativity theory.

2

WAVE EQUATIONS

As a preliminary to the application of wave mechanics to chemical problems it is desirable to acquire an understanding of the characteristics and properties of the wave equations. This chapter will therefore be devoted to establishing some of those concepts associated with wave equations which will be required later.

SIMPLE HARMONIC MOTION AND WAVE EQUATIONS

A particle executes simple harmonic motion if it oscillates about an equilibrium position under the influence of a restoring force which is proportional to the displacement of the particle from the equilibrium position. An example of this type of motion is provided by the vertical movements of a mass suspended from a spring. The motion may be represented as shown in *Figure 2.1*. A point, A, travels round a circular path of radius r with uniform angular velocity ω. If, at zero time, A is at X then at time t, A will have swept out an angle ωt, as shown in *Figure 2.1(a)*. The projection of A on the vertical diameter YOY' is the point B, the displacement of which from the origin is given by

$$y = r \sin \omega t \qquad (2.1)$$

This equation shows that the variation of the displacement with time is represented by a sine wave of amplitude r. This variation is plotted in *Figure 2.1(b)*.

As the point A continues to travel round the circle the displacement of B varies from one extremity of the diameter (OY = $+r$ at $t = \pi/2\omega$) to the other (OY' = $-r$ at $t = 3\pi/2\omega$) and the point B executes simple harmonic motion about the origin O. The motion of B corresponds to the motion of a mass vibrating on a spring

WAVE EQUATIONS 9

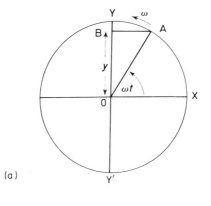

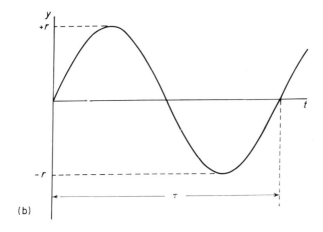

Figure 2.1 Representation of simple harmonic motion

where the origin O corresponds to the equilibrium position of the mass. The motion is periodic in time and is characterized by a *temporal periodicity*, τ, as illustrated in *Figure 2.1(b)*. The time τ is the time required for the point A to make one complete revolution, i.e., the time required for the radius to sweep through an angle of 2π radians. Hence

$$\omega\tau = 2\pi \qquad (2.2)$$

or

$$\omega = 2\pi/\tau \qquad (2.3)$$

The frequency, v_e, of the oscillation of the point B about the

10 WAVE EQUATIONS

equilibrium point is simply the reciprocal of the temporal periodicity

$$v_e = 1/\tau \tag{2.4}$$

Substituting in equation 2.3

$$\boxed{\omega = 2\pi v_e} \tag{2.5}$$

RELATION BETWEEN ENERGY AND AMPLITUDE

The total energy, E, of a particle is the sum of its kinetic energy, T, and its potential energy, V.

$$E = T + V \tag{2.6}$$

For a particle executing simple harmonic motion the total energy must be a constant if there is no damping of the motion, but the kinetic and potential energies vary with time. When the particle is at the equilibrium position represented by the centre of the circle in *Figure 2.1(a)*, its potential energy will be zero ($V = 0$) and T reaches its maximum value, T_{max}, which is equal to E. Thus

$$E = T_{max} \tag{2.7}$$

Now

$$T = \tfrac{1}{2}mv^2 \tag{2.8}$$

where m is the mass of the particle and v its velocity. At the extremities of its oscillation the particle changes direction so that its velocity, and hence T, must be zero. When the particle passes through the equilibrium position its velocity must be a maximum, v_{max}, and the kinetic energy must also have its maximum value, T_{max}, where

Thus,
$$T_{max} = \tfrac{1}{2}mv_{max}^2 \tag{2.9}$$
$$E = \tfrac{1}{2}mv_{max}^2 \tag{2.10}$$

Now v may be expressed as

$$v = \frac{dy}{dt} \tag{2.11}$$

and y is given by equation 2.1 so that

$$v = \frac{d}{dt}[r \sin \omega t] \quad \text{or} \quad v = \omega r \cos \omega t \tag{2.12}$$

v will have a maximum value when $\cos \omega t$ has a maximum value which is, of course, unity. Thus

$$v_{max} = \omega r \tag{2.13}$$

Substituting in equation 2.10

$$E = \tfrac{1}{2}m\omega^2 r^2 \quad (2.14)$$

from which it is seen that *the energy is proportional to the square of the amplitude.* This is an important point to which reference will be made later.

GENERAL DIFFERENTIAL EQUATION OF SIMPLE HARMONIC MOTION

The equation $y = r\sin\omega t$ was obtained by considering the situation shown in *Figure 2.1* when A was at the point X at zero time. That is to say $y = 0$ when $t = 0$. If, at zero time, the point A is at the point Y, i.e., $y = r$ when $t = 0$, the equation representing the motion would have been a cosine function

$$y = r\cos\omega t \quad (2.15)$$

There are thus two equations which can equally well represent simple harmonic motion

$$y_1 = r\sin\omega t \quad (2.16)$$

and

$$y_2 = r\cos\omega t \quad (2.17)$$

The subscripts are appended here simply to distinguish between the two functions. The two equations are both solutions to a differential equation which represents simple harmonic motion. This differential equation may be obtained as follows. The general Newtonian equation of motion is

$$F = ma \quad (2.18)$$

where a is the acceleration of a mass m subjected to a force F. In simple harmonic motion the force is proportional to the displacement but acts towards the origin, i.e., in the opposite direction to the displacement. Thus

$$F = -ky \quad (2.19)$$

where k is a constant known as the force constant. As the velocity of the particle is given by equation 2.11 as $v = dy/dt$ its acceleration will be given by

$$a = \frac{d^2 y}{dt^2} \quad (2.20)$$

Substituting from equations 2.19 and 2.20 into equation 2.18 gives

$$\boxed{m\frac{d^2 y}{dt^2} + ky = 0} \quad (2.21)$$

12 WAVE EQUATIONS

which is the general differential equation of simple harmonic motion to which equations 2.16 and 2.17 are both solutions.

LINEAR COMBINATIONS OF SOLUTIONS

Equation 2.21 is a second order linear homogeneous differential equation. If, to an equation of this type there can be written two solutions, then a linear combination of them is a more general solution. A linear combination of equations 2.16 and 2.17 has the form

$$y = c_1 y_1 + c_2 y_2 \qquad (2.22)$$

where c_1 and c_2 are arbitrary constants, and y would then be a more general solution to the differential equation than either y_1 or y_2. This can be understood by reflecting that equations 2.16 and 2.17 represent special cases. Referring to *Figure 2.1(a)*, the first equation applies when the point A is at X at zero time and the second equation applies when the point A is at Y at zero time. A more general case would be where the point A was somewhere between X and Y at zero time. This more general case is illustrated in *Figure 2.2*. Suppose

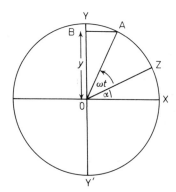

Figure 2.2 The general case of simple harmonic motion

that at zero time the point A was located at Z where the angle ZÔX has the value α. If, at time t, the point A is at the position shown in *Figure 2.2* then the displacement y is given by

$$y = r \sin(\omega t + \alpha) \qquad (2.23)$$

This equation is obviously a more general case of the problem as it allows the point A to be anywhere on the circumference of the circle

WAVE EQUATIONS 13

at zero time. Moreover, equation 2.23 is a linear combination of equations 2.16 and 2.17 as can be understood by expanding equation 2.23.

$$y = r(\sin \omega t \cos \alpha + \cos \omega t \sin \alpha) \quad (2.24)$$

As α is a constant for any particular case, the sine and cosine of α will also be constant. Putting $\cos \alpha = c_1$ and $\sin \alpha = c_2$ and substituting in equation 2.24

$$y = c_1 r \sin \omega t + c_2 r \cos \omega t \quad (2.25)$$

or $y = c_1 y_1 + c_2 y_2$ which is the linear combination illustrated by equation 2.22.

Many of the differential equations encountered in quantum mechanics are of the type of equation 2.21 and it is important to realise that a linear combination of solutions provides a more general solution.

In equation 2.25 r is a constant and this term can be combined with the constants c_1 and c_2 so that the equation may be written in the form

$$y = A \sin \omega t + B \cos \omega t \quad (2.26)$$

which is a general solution to equation 2.21. This can readily be shown by substitution. Equation 2.26 can be differentiated twice

$$\frac{dy}{dt} = \omega A \cos \omega t - \omega B \sin \omega t$$

$$\frac{d^2 y}{dt^2} = -\omega^2 A \sin \omega t - \omega^2 B \cos \omega t$$

$$= -\omega^2 y$$

Writing equation 2.21 in the form

$$\frac{d^2 y}{dt^2} + \frac{k}{m} y = 0 \quad (2.27)$$

Substituting in equation 2.27

$$-\omega^2 y + \frac{k}{m} y = 0$$

This relation may be satisfied if either $y = 0$ or if

$$-\omega^2 + \frac{k}{m} = 0$$

The former is a trivial solution and can only be true when the vibrating particle is at the equilibrium position. The latter condition

14 WAVE EQUATIONS

must provide the general case and thus

$$\omega = (k/m)^{\frac{1}{2}} \tag{2.28}$$

Equation 2.26 or its alternative form, equation 2.23 is thus a solution to equation 2.27 provided that $\omega = (k/m)^{\frac{1}{2}}$.

In addition to expressing the solution to equation 2.27 in the trigonometric form of equation 2.26, the solution may also be expressed in terms of complex exponentials.

(If readers are unfamiliar with complex numbers it is recommended that they read Appendix 1 before proceeding to the next section.)

ALTERNATIVE SOLUTIONS OF THE DIFFERENTIAL EQUATION

Consider the differential equation of simple harmonic motion as in equation 2.27. Putting $k/m = b^2$ the equation takes the form

$$\frac{d^2 y}{dt^2} + b^2 y = 0 \tag{2.29}$$

A solution to this type of equation can be of the form

$$y = K\,e^{qt} \tag{2.30}$$

where K and q are constants. Differentiating equation 2.30 twice gives

$$\frac{dy}{dt} = qK\,e^{qt} = qy \quad \text{and} \quad \frac{d^2 y}{dt^2} = q^2 y$$

Substituting in equation 2.29

$$(q^2 + b^2)y = 0 \tag{2.31}$$

$y = 0$ is a trivial solution which need not be considered further; the alternative, from equation 2.31, is that

$$q^2 + b^2 = 0$$

whence

$$q^2 = -b^2$$

i.e.,

$$q = \pm(-b^2)^{\frac{1}{2}}$$

or

$$q = \pm ib$$

where $i = \sqrt{(-1)}$. Equation 2.30 is thus a solution of equation 2.29 provided q is equal to either ib or $-ib$. There are thus two solutions, y_1 and y_2, where

$$y_1 = K\,e^{ibt}$$

and
$$y_2 = K e^{-ibt}$$

A more general solution would be a linear combination of them both which could be written

$$y = c_1 y_1 + c_2 y_2$$
$$= c_1 K e^{ibt} + c_2 K e^{-ibt}$$

Putting $c_1 K = C$ and $c_2 K = D$ the solution becomes

$$y = C e^{ibt} + D e^{-ibt} \tag{2.32}$$

The constants C and D are completely arbitrary, i.e., equation 2.32 will be a solution of equation 2.29 whatever values are assigned to C and D. This result obtains because equation 2.29 *contains the second derivative of* y *with respect to* t. The solution of a differential equation containing the *first* derivative of a function will contain *one* arbitrary constant; the solution of an equation containing the *second* derivative will contain *two* arbitrary constants, and so on. This can readily be understood from a simple example. Suppose

$$\frac{dy}{dx} = 2 \tag{2.33}$$

Then
$$y = 2x + k \tag{2.34}$$

where k is a constant, which can have any value in equation 2.34 and equation 2.33 will still be true. Similarly, if

$$\frac{d^2 y}{dx^2} = 3 \tag{2.35}$$

then
$$\frac{dy}{dx} = 3x + k_1$$

and
$$y = \tfrac{3}{2} x^2 + k_1 x + k_2 \tag{2.36}$$

The two constants k_1 and k_2 in equation 2.36 can have any values and this equation will still be a solution of equation 2.35. *Values of these constants can only be assigned when some additional limitations are imposed on the problem.*

Returning to equation 2.32 it seems that the displacement, y, is the sum of a real part and an imaginary part as both terms in the equation are complex numbers. It would be an advantage to make y real since it must represent a real displacement and to do this use can be made of the fact that the sum of a complex number and its

16 WAVE EQUATIONS

conjugate is real. Thus, as C and D can have any values put $D = C^*$ where C^* is the conjugate of C. Equation 2.32 thus becomes

$$y = C e^{ibt} + C^* e^{-ibt} \qquad (2.37)$$

where the second term is the conjugate of the first and y must now be entirely real. Further, put

$$C = \frac{r}{2i} e^{i\alpha}$$

so that

$$C^* = -\frac{r}{2i} e^{-i\alpha}$$

where r and α are constants. As C was an arbitrary constant it can be given any value which is convenient. Equation 2.37 thus becomes

$$y = \frac{r}{2i} e^{i\alpha} e^{ibt} - \frac{r}{2i} e^{-i\alpha} e^{-ibt}$$

or

$$y = \frac{r}{2i} \left[e^{i(bt+\alpha)} - e^{-i(bt+\alpha)} \right] \qquad (2.38)$$

Recalling that

$$\sin x = \frac{1}{2i} (e^{ix} - e^{-ix}) \qquad \text{(see Appendix 1)}$$

equation 2.38 may be written

$$y = r \sin(bt + \alpha) \qquad (2.39)$$

This equation shows that the motion is periodic and that

$$b\tau = 2\pi \quad \text{or} \quad b = 2\pi/\tau \qquad (2.40)$$

Comparison of equation 2.40 and equation 2.2 shows that $b = \omega$ and equation 2.39 becomes

$$y = r \sin(\omega t + \alpha)$$

or

$$y = r \sin \omega t \cos \alpha + r \cos \omega t \sin \alpha$$

or, putting $r \cos \alpha = A$ and $r \sin \alpha = B$

$$y = A \sin \omega t + B \cos \omega t$$

which is the trigonometric solution to the differential equation which was given earlier. This trigonometric form may be readily converted

WAVE EQUATIONS 17

to the complex exponential form by using the relations

$$\sin x = \frac{1}{2i}(e^{ix} - e^{-ix})$$

and (see Appendix 1)

$$\cos x = \tfrac{1}{2}(e^{ix} + e^{-ix})$$

and substituting in the trigonometric form so that

$$y = \frac{A}{2i}(e^{i\omega t} - e^{-i\omega t}) + \frac{B}{2}(e^{i\omega t} + e^{-i\omega t})$$

$$= \left(\frac{iB + A}{2i}\right)e^{i\omega t} + \left(\frac{iB - A}{2i}\right)e^{-i\omega t}$$

or, putting

$$\frac{iB + A}{2i} = C \quad \text{and} \quad \frac{iB - A}{2i} = D$$

$$y = C e^{i\omega t} + D e^{-i\omega t}$$

It should thus be remembered that the solution of a differential equation of the form

$$\frac{d^2 y}{dt^2} + \omega^2 y = 0 \qquad (2.41)$$

may be expressed as

$$y = A \sin \omega t + B \cos \omega t \qquad (2.42)$$

or

$$y = C e^{i\omega t} + D e^{-i\omega t} \qquad (2.43)$$

whichever is the more convenient for the problem under consideration.

PROGRESSIVE WAVES

Taking one of the simple equations for simple harmonic motion

$$y = r \sin \omega t$$

and remembering that $\omega = 2\pi/\tau$ the equation may be written

$$y = r \sin \frac{2\pi}{\tau} t \qquad (2.44)$$

This equation shows how the displacement varies with time; it is a *temporally dependent* equation. If the mass oscillating on the end of the spring were immersed in a fluid it would set up a train of

18 WAVE EQUATIONS

progressive waves on the surface which would be both *temporally and spatially dependent*. If the wave were photographed at some instant in time (say zero time) it would have the shape of a sine wave and a plot of displacement, φ, against distance, x, from the source would be sinusoidal as illustrated in *Figure 2.3*. This wave has a spatial periodicity, λ, which is the wavelength.

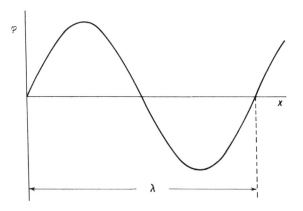

Figure 2.3 Progressive waves at zero time

When the displacement varies with time with a temporal periodicity τ, the equation is 2.44. Thus, by analogy, when displacement varies with distance with a spatial periodicity λ the appropriate equation should be

$$\varphi = r \sin \frac{2\pi}{\lambda} x \qquad (2.45)$$

at zero time. At some later time, t, the wave as a whole will have moved forward a distance vt where v is the velocity of the wave. This situation is illustrated in *Figure 2.4*. The shape of the wave will still be represented by a sine function if a distance vt is subtracted from every value of x, hence

$$\varphi = r \sin \frac{2\pi}{\lambda} (x - vt) \qquad (2.46)$$

gives the dependence of displacement on distance at any time t.

Equation 2.46 is one solution to the general differential equation of wave motion which applies to any wave of any shape. If the shape is unknown then φ will be some function of $(x - vt)$, i.e.,

$$\varphi = f(x - vt)$$

WAVE EQUATIONS

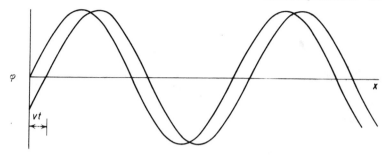

Figure 2.4 *Progressive waves at time t*

The arbitrary function may be eliminated by differentiating twice with respect to each variable.

$$\left(\frac{\partial \varphi}{\partial x}\right)_t = f'(x-vt)$$

$$\left(\frac{\partial^2 \varphi}{\partial x^2}\right)_t = f''(x-vt) \qquad (2.47)$$

and

$$\left(\frac{\partial \varphi}{\partial t}\right)_x = -vf'(x-vt)$$

$$\left(\frac{\partial^2 \varphi}{\partial t^2}\right)_x = v^2 f''(x-vt) \qquad (2.48)$$

Comparison of equations 2.47 and 2.48 shows that

$$\frac{\partial^2 \varphi}{\partial x^2} = \frac{1}{v^2}\frac{\partial^2 \varphi}{\partial t^2} \qquad (2.49)$$

which is the general differential wave equation.

STANDING WAVES

The wave which has been considered in *Figure 2.4*, given by equation 2.46, applies only to motion in one dimension and is better considered as applying to the vibration of a string. If the string is fixed at the end the wave may be reflected from the fixed point and travel back down the string. As its direction of motion is reversed, the velocity of the reflected wave is $-v$ and its equation is given by

20 WAVE EQUATIONS

replacing v by $-v$ in equation 2.46 to give

$$\varphi = r \sin \frac{2\pi}{\lambda}(x+vt) \tag{2.50}$$

If the string is fixed at both ends the progressive waves given by equations 2.46 and 2.50 will travel up and down the string simultaneously and the net displacement of the string from its equilibrium position will be given by the sum of the two individual displacements as

$$\varphi = r \sin \frac{2\pi}{\lambda}(x-vt) + r \sin \frac{2\pi}{\lambda}(x+vt) \tag{2.51}$$

Now, for any progressive wave motion, $v = v\lambda$, where v is the frequency of oscillation of the wave, so that equation 2.51 may be written in the form

$$\varphi = r \sin \frac{2\pi}{\lambda}(x-v\lambda t) + r \sin \frac{2\pi}{\lambda}(x+v\lambda t)$$

or

$$\varphi = r \sin 2\pi \left(\frac{x}{\lambda} - vt\right) + r \sin 2\pi \left(\frac{x}{\lambda} + vt\right) \tag{2.52}$$

Expanding the trigonometric terms in equation 2.52

$$\varphi = r \left(\sin \frac{2\pi}{\lambda} x \cos 2\pi vt - \cos \frac{2\pi}{\lambda} x \sin 2\pi vt \right)$$
$$+ r \left(\sin \frac{2\pi}{\lambda} x \cos 2\pi vt + \cos \frac{2\pi}{\lambda} x \sin 2\pi vt \right)$$

or

$$\varphi = 2r \sin \frac{2\pi}{\lambda} x \cdot \cos 2\pi vt \tag{2.53}$$

Inspection of equation 2.53 shows that φ will be zero whenever $\sin(2\pi/\lambda)x$ is zero, regardless of the value of t. Now

$$\sin \frac{2\pi}{\lambda} x = 0$$

whenever

$$\frac{2\pi}{\lambda} x = n\pi$$

where n is an integer. That is, whenever $x = n(\lambda/2)$.

Such a wave represented by equation 2.53 is called a *standing wave* and it will be realised that these waves are naturally associated with a set of integers. In fixing the string at each end 'boundary conditions' have been imposed on the situation and it was Schrödinger who realised that the imposition of boundary conditions on the flooded planet model would generate automatically a set of integers.

The solution of the wave equation obtained above for a standing wave, equation 2.53, was obtained by considering the progressive wave as

$$\varphi = r \sin \frac{2\pi}{\lambda}(x - vt) \tag{2.46}$$

If the starting point had been the progressive wave equation

$$\varphi = r \cos \frac{2\pi}{\lambda}(x - vt) \tag{2.54}$$

the solution for the standing wave would have been

$$\varphi = 2r \cos \frac{2\pi}{\lambda} x . \cos 2\pi vt \tag{2.55}$$

Two solutions for standing waves are thus given by equations 2.53 and 2.55. These equations may be written

$$\varphi_1 = 2r \sin \frac{2\pi}{\lambda} x . \cos 2\pi vt \tag{2.56}$$

and

$$\varphi_2 = 2r \cos \frac{2\pi}{\lambda} x . \cos 2\pi vt \tag{2.57}$$

A more general solution for standing waves would result from a linear combination of equations 2.56 and 2.57,

$$\varphi = A\varphi_1 + B\varphi_2 \tag{2.58}$$

where A and B are arbitrary constants. The linear combination may thus be written

$$\varphi = A . 2r \sin \frac{2\pi}{\lambda} x . \cos 2\pi vt + B \ 2r \cos \frac{2\pi}{\lambda} x . \cos 2\pi vt$$

or,

$$\varphi = [2r \cos 2\pi vt] \left[A \sin \frac{2\pi}{\lambda} x + B \cos \frac{2\pi}{\lambda} x \right] \tag{2.59}$$

This equation may be expressed alternatively in terms of complex exponentials by putting

$$A = k \sin \alpha \quad \text{and} \quad B = k \cos \alpha$$

22 WAVE EQUATIONS

when

$$\varphi = [2r \cos 2\pi v t] \left[k \cos\left(\frac{2\pi}{\lambda} x - \alpha\right) \right]$$

$$\varphi = [2r \cos 2\pi v t] \left\{ \tfrac{1}{2}k \exp\left[i\left(\frac{2\pi}{\lambda} x - \alpha\right) \right] + \tfrac{1}{2}k \exp\left[-i\left(\frac{2\pi}{\lambda} x - \alpha\right) \right] \right\}$$

$$\varphi = [2r \cos 2\pi v t] \left[\tfrac{1}{2}k \exp(-i\alpha) \cdot \exp\left(\frac{2\pi i}{\lambda} x\right) \right.$$

$$\left. + \tfrac{1}{2}k \exp(i\alpha) \cdot \exp\left(-\frac{2\pi i}{\lambda} x\right) \right] \quad (2.60)$$

Putting $\tfrac{1}{2}k\,e^{-i\alpha} = C$ and $\tfrac{1}{2}k\,e^{i\alpha} = D$, equation 2.60 becomes

$$\varphi = [2r \cos 2\pi v t] \left[C \exp\left(\frac{2\pi i}{\lambda} x\right) + D \exp\left(-\frac{2\pi i}{\lambda} x\right) \right] \quad (2.61)$$

Equations 2.59 and 2.61 are thus solutions of the differential wave equation for standing waves. In both these equations there is a bracket which describes how φ varies with x, the space part, and a bracket which describes how φ varies with t, the time part. In quantum mechanics the time part is important when magnetic properties associated with the movement of bound electrons are being considered but in most cases the interest lies in the average value of φ over an appreciable interval of time and how this value of φ depends on x. Under these conditions it is legitimate to consider only the space part of the wave function.

THE SCHRÖDINGER WAVE EQUATION

If it is only necessary to consider the space part of the wave equation, the time part may be eliminated by differentiation. Consider one of the simple solutions to the differential wave equation

$$\frac{\partial^2 \varphi}{\partial x^2} = \frac{1}{v^2} \frac{\partial^2 \varphi}{\partial t^2} \quad (2.49)$$

which is

$$\varphi = 2r \sin \frac{2\pi}{\lambda} x \cdot \cos 2\pi v t \quad (2.53)$$

for standing waves. Writing equation 2.53 as

$$\varphi = \psi(x) \cdot \cos 2\pi v t$$

WAVE EQUATIONS 23

where $\psi(x)$ is the function of x, and differentiating

$$\frac{\partial \varphi}{\partial x} = \psi'(x) \cdot \cos 2\pi vt$$
$$\frac{\partial^2 \varphi}{\partial x^2} = \psi''(x) \cdot \cos 2\pi vt \tag{2.62}$$

and

$$\frac{\partial \varphi}{\partial t} = -\psi(x) \cdot 2\pi v \sin 2\pi vt$$
$$\frac{\partial^2 \varphi}{\partial t^2} = -\psi(x) \cdot 4\pi^2 v^2 \cos 2\pi vt \tag{2.63}$$

Substituting from equations 2.62 and 2.63 into equation 2.49

$$\psi''(x) \cdot \cos 2\pi vt = \frac{1}{v^2}[-\psi(x) \cdot 4\pi^2 v^2 \cos 2\pi vt]$$

or

$$\psi''(x) = -\frac{4\pi^2 v^2}{v^2}\psi(x) \tag{2.64}$$

As $v/\nu = \lambda$, equation 2.64 may be written

$$\psi''(x) = -\frac{4\pi^2}{\lambda^2}\psi(x)$$

or

$$\psi''(x) + \frac{4\pi^2}{\lambda^2}\psi(x) = 0 \tag{2.65}$$

The above process has thus eliminated the time variable in equation 2.65 which may also be written in the alternative form

$$\frac{d^2\psi}{dx^2} + \frac{4\pi^2}{\lambda^2}\psi = 0 \tag{2.66}$$

To apply this equation to particles, λ must be replaced by a momentum term using the de Broglie relationship

$$\lambda = h/p \tag{2.67}$$

where h is Planck's constant and p is the momentum of the particle. From equation 2.67

$$\frac{1}{\lambda^2} = \frac{p^2}{h^2} = \frac{m^2 v^2}{h^2}$$

Substituting in equation 2.66

$$\frac{d^2\psi}{dx^2} + \frac{4\pi^2 m^2 v^2}{h^2}\psi = 0 \tag{2.68}$$

24 WAVE EQUATIONS

It will be remembered that $T = \frac{1}{2}mv^2$ so that

$$v^2 = \frac{2T}{m} \tag{2.69}$$

Furthermore, $T = E - V$ so that equation 2.69 may be written $v^2 = 2/m(E-V)$. Substituting in equation 2.68

$$\frac{d^2\psi}{dx^2} + \frac{8\pi^2 m}{h^2}(E-V)\psi = 0 \tag{2.70}$$

This equation is the time independent Schrödinger equation in one dimension. For three dimensions it becomes

$$\frac{\partial^2\psi}{\partial x^2} + \frac{\partial^2\psi}{\partial y^2} + \frac{\partial^2\psi}{\partial z^2} + \frac{8\pi^2 m}{h^2}(E-V)\psi = 0$$

or, more briefly

$$\boxed{\nabla^2\psi + \frac{8\pi^2 m}{h^2}(E-V)\psi = 0} \tag{2.71}$$

where ∇^2 is known as the Laplacian operator. The Schrödinger wave equation is an equation which describes the properties of particles in terms of wave motion.

EIGENFUNCTIONS AND EIGENVALUES

When the Schrödinger equation is applied to a particular problem the differential equation 2.71 must be solved. There will be many expressions for ψ which satisfy the Schrödinger equation, but only some of the solutions will be acceptable for the problem under consideration. The acceptable solutions are called *eigenfunctions*, or well-behaved functions, and the corresponding energies, E, are called the *eigenvalues* of the system.

The type of conditions which a wave function must satisfy before it is acceptable will be discussed in the next chapter but the principle can be illustrated here by a simple example. Consider the differential equation

$$\frac{dy}{dx} = 2 \tag{2.72}$$

The solution to this equation is

$$y = 2x + k \tag{2.73}$$

WAVE EQUATIONS 25

where k is an arbitrary constant which can have any value. No matter what the value of k, differentiation of equation 2.73 will always lead to equation 2.72. There are thus an infinite number of solutions to equation 2.72, each solution being a straight line of slope 2. If however, the condition is imposed, that the solution to equation 2.72 must pass through the point $x = 1$, $y = 5$, it will be seen that there is only one value of k which will fulfil this condition and that is $k = 3$. The only acceptable solution to equation 2.72 in view of the restriction imposed by the condition is thus

$$y = 2x + 3$$

In the above example only one arbitrary constant was involved, but with the Schrödinger equation, a second order differential equation, there will be two arbitrary constants involved.

In view of the material discussed in the earlier part of this chapter the general solutions to the Schrödinger equation may be written for those cases where the potential energy of the system is constant. Taking the one-dimensional Schrödinger equation for simplicity

$$\frac{d^2\psi}{dx^2} + \frac{8\pi^2 m}{h^2}(E-V)\psi = 0 \qquad (2.70)$$

If the potential energy, V, is constant and for a particular state of the system the total energy, E, is constant then

$$\frac{8\pi^2 m}{h^2}(E-V) = k^2$$

where k is a constant and equation 2.70 may be written

$$\boxed{\frac{d^2\psi}{dx^2} + k^2\psi = 0} \qquad (2.74)$$

An equation of this type has been solved earlier in this chapter. The equation was equation 2.41 and the solutions in terms of trigonometric functions and complex exponentials were given in equations 2.42 and 2.43 respectively. By analogy then, the solutions of the Schrödinger equation for constant V may be written as

$$\boxed{\psi = A \sin kx + B \cos kx} \qquad (2.75)$$

or

$$\boxed{\psi = C e^{ikx} + D e^{-ikx}} \qquad (2.76)$$

where A and B or C and D are arbitrary constants. Those solutions which are eigenfunctions will be obtained by imposing boundary conditions on the problems considered to arrive at acceptable values for the constants.

INTERPRETATION OF ψ

When the wave equation is solved to obtain a wave function, ψ, it may turn out to be a complex function. Neither ψ nor its conjugate ψ^*, has any physical significance. The product of a complex number and its conjugate is a real quantity however, so that the product $\psi\psi^*$ is a real function of the co-ordinates and some physical significance may be attributed to it.

It will be remembered that the energy of a wave is proportional to the square of the amplitude (see equation 2.14); moreover the energy of a wave disturbance is a measure of its intensity and thus intensity is proportional to the square of the amplitude. If, now, Thomson and Reid's work on the diffraction of electrons is considered it will be recalled that a beam of electrons may be diffracted by a crystal and if allowed to fall on a fluorescent screen will show bright patterns at the points where most electrons fall. If only one electron were considered it might fall anywhere on the screen, but there must be a higher probability of its landing in the bright parts where the intensity of fluorescence is greater. The probability of finding an electron at a particular point may thus be related to the intensity of fluorescence and hence to the square of the amplitude of the wave representing the behaviour of the electron. As the wave function has the nature of an amplitude the probability of finding the electron at a particular point may thus be related to the value of $\psi\psi^*$ at that point.

$\psi\psi^*$ is usually interpreted as being proportional to the probability of finding a particle at a particular point. $\psi\psi^* dV$ can thus be regarded as proportional to the probability of finding a particle in an element of volume dV or, alternatively, it may be regarded as proportional to the particle density in the volume dV.

This interpretation stems from the fact that electrons have some properties for which a particle description is best suited and some properties requiring description by wave equations. The possession of both particle and wave properties is one aspect of a fundamental postulate of modern physics adduced by Heisenberg in 1927. This is the Uncertainty Principle which states that the simultaneous exact determination of the momentum (or energy) of a particle and its position is impossible. When an electron exhibits particle properties its position can be known with some exactness, but there will be an uncertainty with regard to its momentum. When it exhibits wave properties, as in diffraction, the momentum can be specified but the position is uncertain. In fact, denoting the uncertainty in momentum as Δp and the uncertainty in position as Δx, there exists the relationship

$$\Delta p . \Delta x \approx h$$

Since the position of an electron of definite energy cannot be known exactly, the quantity $\psi\psi^*$ is interpreted in terms of the probability of the electron's position.

The points discussed in this chapter are illustrated by the application of the Schrödinger equation to some simple quantum mechanical problems in the next chapter.

3

PARTICLE IN A BOX

There are only a few situations in which an exact solution of the Schrödinger equation may be obtained. Some of these are hypothetical but they approximate to real situations and also serve the purpose of illustrating the application of the Schrödinger equation. In this chapter the application of the Schrödinger equation to a hypothetical problem is considered and some basic concepts of quantum mechanics are introduced.

PARTICLE IN A ONE-DIMENSIONAL BOX

Suppose a particle of mass m is restricted to motion in one dimension, along the x-axis say. Suppose that between $x = 0$ and $x = a$ the potential energy is zero and that outside this range the potential is infinite. A plot of potential energy against distance under these circumstances would have the appearance of *Figure 3.1*. Outside the limits $x = 0$ to $x = a$, the particle would have an infinite potential energy. This is clearly impossible and the probability of finding the particle outside this restricted region is zero. For this reason the problem is sometimes called the particle in a box problem. As the particle can only be inside the box $\psi\psi^*$ and hence ψ must be zero when $0 \geqslant x \geqslant a$.

Inside the box $V = 0$ and the one-dimensional form of the Schrödinger equation, equation 2.70 becomes

$$\frac{d^2\psi}{dx^2} + \frac{8\pi^2 m}{h^2} E\psi = 0 \qquad (3.1)$$

From equation 2.75 the solution of equation 3.1 may be written as

$$\psi = A \sin kx + B \cos kx \qquad (3.2)$$

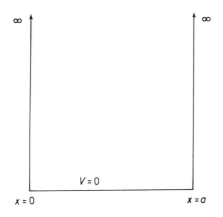

Figure 3.1 One-dimensional potential box

where
$$k^2 = 8\pi^2 mE/h^2 \tag{3.3}$$

Equation 3.2 is the general solution to equation 3.1 where A and B are arbitrary constants but, in order to obtain an eigenfunction of the system, limitations must be imposed on equation 3.2 and these stem from the boundary conditions of the problem. For example, it is known that when $x = 0$, $\psi = 0$. Putting $x = 0$ in equation (3.2) yields

$$\psi = A \sin 0 + B \cos 0$$

As $\sin 0 = 0$ and $\cos 0 = 1$, $\psi = B$. The particular problem, however, requires that $\psi = 0$ when $x = 0$ and from this it is clear that the value of B must be zero. Equation 3.2 thus reduces to

$$\psi = A \sin kx \tag{3.4}$$

The other boundary condition of the problem is that $\psi = 0$ when $x = a$ and substitution of these values in equation 3.4 gives

$$A \sin ka = 0 \tag{3.5}$$

The sine of an angle is only zero when the angle itself is zero or equal to an integral multiple of π radians. If the angle itself were zero then ψ would be zero everywhere inside the box and this is not acceptable as the particle must be somewhere in the box. It follows from equation 3.5 therefore, that $ka = n\pi$ where $n = 1, 2, 3, \ldots$.

Hence $k = n\pi/a$ is a necessary condition for the solution of the wave equation to be acceptable, i.e., to be an eigenfunction of the

30 PARTICLE IN A BOX

system. Applying this condition to equation 3.4, the eigenfunctions, ψ_n, of the system are given by

$$\psi_n = A \sin \frac{n\pi}{a} x \qquad (3.6)$$

The total energy of the particle may now be determined. Equation 3.3 can be written as

$$E = k^2 h^2 / 8\pi^2 m$$

and substituting $n\pi/a$ for k the eigenvalues, E_n, are

$$E_n = \frac{n^2 h^2}{8ma^2} \qquad (3.7)$$

The integer n may be termed a quantum number and it is to be noted that the energy of the particle cannot have a continuous range of values but is quantized in units of $h^2/8ma^2$, the multiples being $1, 4, 9, 16, \ldots, n^2$.

It has been pointed out above that n cannot be zero. As a consequence the energy of the particle can never be zero, the minimum value being $h^2/8ma^2$ when $n = 1$. This is known as the *zero point energy* of the system and is a feature of all systems where the motion is vibratory. The particle in the box is a vibratory system as it can move backwards and forwards along the x-axis between the limits $x = 0$ and $x = a$.

The energy level diagram of the particle is shown in *Figure 3.2* for the first four levels. The eigenfunctions (ψ) and probability distributions (ψ^2) are also plotted in *Figure 3.2* where the appropriate energy levels are used as the x-axes for these functions.

For the lowest energy state half a wavelength of the wave function fits into the box and there is no point inside the box where the amplitude of the wave function is zero. Such a point would be called a *node* and it can be seen from *Figure 3.2* that each eigenfunction has $(n-1)$ internal nodes. The greater the number of nodes, the greater is the kinetic energy (in this particular case, the total energy since $V = 0$) of the system.

For a given value of the quantum number the energy of the particle is inversely proportional to the mass of the particle and the square of the length of the box. Further, it will be appreciated that as the particle becomes more massive or the box longer, the energy levels become more closely spaced. With particles of large mass able to move over large distances the energy levels become so closely spaced as to seem to allow a continuous range of energies. This is the result that classical mechanics would give for an analogous problem on a large scale. For example, a tennis ball bouncing backwards and forwards between two walls could have any energy. Quantum

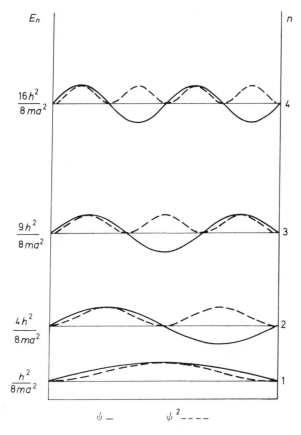

Figure 3.2 *Energy level diagram with eigenfunctions and probability distributions*

mechanics thus gives the same result as classical mechanics for systems of dimensions such that $8ma^2 \gg n^2h^2$. This is an illustration of the *correspondence principle* which states that in the limit when the quantum numbers describing the system become very large the quantum mechanical result must become identical with the classical result.

32 PARTICLE IN A BOX

NORMALISATION OF WAVE FUNCTIONS

It was pointed out in the last chapter that the probability of finding the particle in an element dx is proportional to $\psi\psi^* dx$. In the case of the particle in a box the eigenfunctions given by equation 3.6 do not contain $i[=\sqrt{(-1)}]$, so that the conjugate of ψ is the same as ψ and

$$\psi\psi^* dx = \psi^2 dx$$

It would be useful if $\psi^2 dx$ were not merely *proportional* to the probability but actually *equal* to it. In the wave equations given by equation 3.6

$$\psi_n = A \sin \frac{n\pi}{a} x \tag{3.6}$$

the constant A is an arbitrary constant and its magnitude may be chosen to fit any condition provided the eigenvalues are unaffected. The value of A may be chosen so that $\psi^2 dx$ is *equal* to the probability of finding the particle in the element dx.

The particle must be somewhere between the limits $x = 0$ and $x = a$ so that the probability of finding it between these finite limits must be unity. Hence

$$\boxed{\int_0^a \psi^2 dx = 1} \tag{3.8}$$

Imposing this condition, with the attendant consequences for the values of any constants in the wave equation is called *normalisation* of the wave equation.

Equation 3.8 in the more general case takes the form

$$\int \psi\psi^* d\tau = 1$$

where the integral implies integration over the whole of the space where the particle may exist.

For the particle in the box equation 3.8 becomes

$$\int_0^a \left[A \sin \frac{n\pi}{a} x \right]^2 dx = 1$$

or

$$A^2 \int_0^a \left[\sin^2 \frac{n\pi}{a} x \right] dx = 1 \tag{3.9}$$

Since $\sin^2 \alpha = \frac{1}{2}(1 - \cos 2\alpha)$ equation 3.9 may be written

$$\tfrac{1}{2}A^2 \int_0^a \left(1 - \cos \frac{2n\pi}{a} x\right) dx = 1$$

whence

$$\tfrac{1}{2}A^2 \left[x - \frac{a}{2n\pi} \sin \frac{2n\pi}{a} x \right]_0^a = 1$$

$$\tfrac{1}{2}A^2 \left[a - \frac{a}{2n\pi} \sin 2n\pi \right] = 1$$

Remembering that n is an integer, the second term in the bracket above contains the sine of angles that are integral multiples of 2π radians. Any angle which is such a multiple has a sine of zero so that the above expression becomes

$$\tfrac{1}{2}A^2 a = 1$$

or

$$A = \sqrt{\frac{2}{a}}$$

The normalised eigenfunctions are thus given by

$$\boxed{\psi_n = \sqrt{\left(\frac{2}{a}\right)} \sin \frac{n\pi}{a} x} \quad (3.10)$$

and the probability of finding the particle at a particular value of x is given by

$$\psi_n^2 = \frac{2}{a} \sin^2 \frac{n\pi}{a} x \quad (3.11)$$

It can be seen from *Figure 3.2* that the probability of finding the particle at a node is zero and equation 3.11 shows that the probability has a maximum value of $2/a$ between the nodes. It should be noted that for the lowest energy level the maximum probability occurs at the mid-point of travel of the particle. Further, as the quantum number becomes very large the points of maximum probability become more numerous and hence closer together. As the distances between points of maximum probability become too small to measure it will appear that the probability is uniform over the length of the box which is the classical result.

ORTHOGONALITY

It is a mathematical property of eigenfunctions that if any two of them are exact solutions of the wave equation then they are orthogonal. This means that they are independent of one another and the integral of their product over the whole of the space in which the system exists is equal to zero. For the general case

$$\int \psi_n \psi_m \, d\tau = 0 \quad (n \neq m)$$

or for the specific case of the particle in the box

$$\int_0^a \psi_n \psi_m \, dx = 0 \quad (n \neq m)$$

where n and m are different values of the quantum number. The orthogonality of a pair of eigenfunctions for the particle in a box

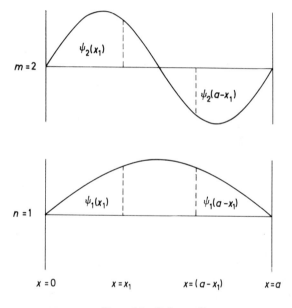

Figure 3.3 Orthogonality

can easily be seen from a diagram of the wave functions in cases where one of the quantum numbers is an even number and the other is an odd number. Figure 3.3 illustrates the eigenfunctions for $n = 1$ and $m = 2$. Consider the values of ψ_1 and ψ_2 at two values of x which are symmetrically located about the mid-point of the box. These values of x are denoted $x = x_1$ and $x = (a - x_1)$. Now,

$$\psi_2(x_1) = -\psi_2(a - x_1)$$

and
$$\psi_1(x_1) = \psi_1(a-x_1)$$

It is thus clear that
$$\psi_2(x_1)\cdot\psi_1(x_1) = -\psi_2(a-x_1)\cdot\psi_1(a-x_1)$$

The integral of the product $\psi_1\cdot\psi_2$ over all values of x is simply the sum of every $\psi_1\psi_2$ term at all values of x and from the above considerations it may be understood that the sum of every $\psi_1\psi_2$ term for values of x between $x=0$ and $x=a/2$ (the left-hand side of the box) is equal and opposite to the sum of every $\psi_1\psi_2$ term for values of x between $x=a/2$ and $x=a$ (the right-hand side of the box). The integral of the product from $x=0$ to $x=a$ must therefore be zero.

The condition of orthogonality cannot readily be understood in diagrammatic terms when both n and m are even or both are odd, but the proposition may be proved for any values of n and m.

$$\int_0^a \psi_n\psi_m\,dx = \int_0^a \left[\sqrt{\left(\frac{2}{a}\right)}\sin\frac{n\pi}{a}x\right]\left[\sqrt{\left(\frac{2}{a}\right)}\sin\frac{m\pi}{a}x\right]dx$$

$$= \frac{2}{a}\int_0^a \left(\sin\frac{n\pi}{a}x\right)\left(\sin\frac{m\pi}{a}x\right)dx$$

Since $2\sin\alpha\sin\beta = \cos(\alpha-\beta)-\cos(\alpha+\beta)$ the integral may be written as

$$\int_0^a \psi_n\psi_m\,dx$$

$$= \frac{2}{a}\int_0^a \frac{1}{2}\left\{\cos\left[(n-m)\frac{\pi}{a}x\right]-\cos\left[(n+m)\frac{\pi}{a}x\right]\right\}dx$$

$$= \frac{1}{a}\left\{\frac{a}{(n-m)\pi}\sin\left[(n-m)\frac{\pi}{a}x\right]-\frac{a}{(n+m)\pi}\sin\left[(n+m)\frac{\pi}{a}x\right]\right\}_0^a$$

$$= \frac{1}{a}\left\{\frac{a}{(n-m)\pi}\sin[(n-m)\pi]-\frac{a}{(n+m)\pi}\sin[(n+m)\pi]\right\}$$

Since the difference and the sum of the two quantum numbers n and m must be integers, the sines of the angles are all zero and hence the integral is zero.

Wave functions which are both normalised and orthogonal are said to be *orthonormal*.

It was stated above that only eigenfunctions which are exact solutions to the wave equation form orthogonal pairs. In many chemical problems only approximate solutions are obtained and these are not necessarily orthogonal. In this case they must be made orthogonal or allowance may have to be made for the fact that they are not.

QUANTUM MECHANICAL OPERATORS

Before considering quantum mechanical operators it is probably best to clarify the meaning of the term operator in the mathematical sense. An operator conveys the instruction to carry out an operation on a function. Thus the symbol $\sqrt{}$ is an operator, the instruction being to take the square root of the quantity which follows it. For example

$$\sqrt{x^4} = x^2$$

Operators always operate on the functions written to the right of them. A further example of a mathematical operation is differentiation. The symbol d/dx is an operator which instructs that the following function should be differentiated with respect to x. Thus

$$\frac{d}{dx} x^2 = 2x$$

Multiplication may be regarded as an operation so that the operator x. means multiply the following function by x. Thus

$$x \cdot x = x^2$$

Operators may occur in combinations such as the operator $x \cdot (d/dx)$ which instructs that the following function should first be differentiated with respect to x and then multiplied by x. Thus

$$x \cdot \frac{d}{dx} x^4 = x \cdot 4x^3 = 4x^4$$

With combined operations, the operators are used, starting with the right-hand one.

It should be noted that for the differential operator

$$\frac{d}{dx}(x^5 + 2x^3) = \frac{d}{dx} x^5 + \frac{d}{dx} 2x^3$$

Moreover,

$$\frac{d}{dx}(4x^2) = 4 \frac{d}{dx}(x^2)$$

An operator which fulfils these conditions is known as a *linear operator*.

The operator $\sqrt{}$ is not a linear operator since
$$\sqrt{(16+25)} \neq \sqrt{(16)} + \sqrt{(25)}$$
Having clarified the term operator, consider the one-dimensional Schrödinger equation
$$\frac{d^2\psi}{dx^2} + \frac{8\pi^2 m}{h^2}(E-V)\psi = 0$$
which may be written as
$$-\frac{h^2}{8\pi^2 m} \cdot \frac{d^2\psi}{dx^2} + V\psi = E\psi$$
or
$$\left[-\frac{h^2}{8\pi^2 m} \cdot \frac{d^2}{dx^2} + V\right]\psi = E\psi \qquad (3.12)$$

It will be recalled that the total energy, E, of a system is the sum of the kinetic energy, T, and the potential energy, V.
$$T + V = E \qquad (3.13)$$

Comparing equations 3.12 and 3.13 it may be considered that there is some correspondence between the first term in the bracket of equation 3.12 and the kinetic energy of the system. The kinetic energy is given by
$$T = \tfrac{1}{2}mv^2 = \frac{p^2}{2m}$$
where p is the momentum, so that equation 3.13 may be written in the form
$$\frac{p^2}{2m} + V = E \qquad (3.14)$$

Comparing equations 3.12 and 3.14 it seems that p^2 in equation 3.14 has been replaced by the operator
$$-\frac{h^2}{4\pi^2} \cdot \frac{d^2}{dx^2}$$
in equation 3.12.

Alternatively it may be said that p has been replaced by the operator
$$\frac{h}{2\pi i} \cdot \frac{d}{dx}$$
since
$$p^2 = p \cdot p$$

38 PARTICLE IN A BOX

and

$$\frac{h}{2\pi i} \cdot \frac{d}{dx}\left(\frac{h}{2\pi i} \cdot \frac{d}{dx}\right) = -\frac{h}{4\pi^2} \cdot \frac{d^2}{dx^2}$$

This replacement of a physically observable quantity by an operator is part of the general theory of quantum mechanics and is embodied in the postulatory approach to quantum mechanics which is considered in more detail below.

THE POSTULATES OF QUANTUM MECHANICS

In this book the Schrödinger equation was obtained from the classical relationship for an harmonic standing wave and the de Broglie equation. This approach is useful in illustrating some of the basic concepts, but in general the foundations of quantum mechanics are best considered to be a set of postulates from which the equations of motion may be derived. The original postulates are then justified by the fact that the solutions of these equations agree with experiment.

This axiomatic approach is probably familiar to the reader in other contexts, but it might be useful to quote a simple example here to illustrate the point.

It is convenient to denote $x.x.x.x.x.x$ as x^6 and it is therefore *postulated* that

$$x^6 = x.x.x.x.x.x$$

In general, x^n means x multiplied by itself n times. The notation may be extended by considering

$$\frac{x.x.x.x.x.x}{x.x.x.x} = x.x$$

or, using the above postulate

$$\frac{x^6}{x^4} = x^2$$

It is possible to generalise by writing

$$\frac{x^m}{x^n} = x^{m-n} \qquad (3.15)$$

provided that $m > n$. If the rule embodied in equation 3.15 is applied when $m = n$ it gives

$$\frac{x^m}{x^m} = x^0 \qquad (3.16)$$

PARTICLE IN A BOX   39

This result is not very meaningful as it is difficult to understand the significance of the right-hand side of equation 3.16 which means x multiplied by itself zero times. In fact, it is clear that

$$\frac{x^m}{x^m} = 1$$

and for equation 3.15 to apply a *further postulate* is required, viz.,

$$x^0 = 1$$

By means of this second postulate equation 3.15 is now true for $m \geq n$.

If equation 3.15 is applied to situations where $m < n$, another meaningless result is obtained. For example

$$\frac{x^2}{x^5} = x^{-3}$$

The right-hand side of this equation means x multiplied by itself -3 times. The significance of this statement is not at all clear. In fact,

$$\frac{x^2}{x^5} = \frac{1}{x^3}$$

and for equation 3.15 to be generally applicable a further postulate is required, viz.,

$$x^{-n} = \frac{1}{x^n}$$

With this additional qualification equation 3.15 holds for any values of m and n. Thus, in order to deal with operations involving powers of numbers, a set of three postulates are taken which enable various operations to be performed with the help of equation 3.15 to generate self-consistent results.

The postulates of quantum mechanics are similar in principle. They consist of a set of rules which, if followed, lead to self-consistent results. The postulates of quantum mechanics are a little more complex than in the example given above, but they are considered below and some discussion of the postulates should help to clarify their operation. For the purposes of the material covered in this book two postulates may be stated.

Postulate 1

A quantum mechanical system of n particles is described as completely as possible by a function, $\psi(q_1, q_2, \ldots, q_{3n}, t)$, called

40 PARTICLE IN A BOX

the wave function, which determines all measurable quantities of the system.

The wave function ψ is required to be a finite, single-valued, continuous function which becomes zero at infinity. It must have an integrable square and its first derivatives $\partial\psi/\partial q_i$ must also be continuous. ψ is interpreted physically by $\psi\psi^* dq_1, \ldots, dq_{3n}$ being the probability of finding the particles with co-ordinates lying between $q_1, \ldots, q_{3n}$ and $q_1 + dq_1, \ldots, q_{3n} + dq_{3n}$.

Since each particle must be somewhere in space, the integrated probability density must be unity, i.e.,

$$\int \psi\psi^* d\tau = 1$$

This first postulate mostly covers the material which has been considered up to this point in rather formal terms and it includes some extra conditions which a wave function must fulfil if it is to be an eigenfunction of the system.

For the single particle in a one-dimensional box only one co-ordinate, the x co-ordinate is required so that for this particular problem the wave function was a function of only one co-ordinate $\psi(x)$. If the particle was free to move in three dimensions then three co-ordinates would have been required. For a system of n particles then, $3n$ co-ordinates are required and hence ψ will be a function of $(q_1, q_2, \ldots, q_{3n})$ where the symbol q represents a co-ordinate. For the material in this book the time co-ordinate, t, is not included because only the time independent Schrödinger equation is considered.

If $\psi\psi^*$ is to be interpreted as a probability it is reasonable to require that ψ should be finite, otherwise there could be an infinite probability of finding the particle somewhere. It is further reasonable to specify that it should be single valued, otherwise there might be two or more probabilities of finding a particle at a particular point at the same time and this would not be meaningful. It is also difficult to understand how a particle can be at infinity so that the wave function is required to become zero at infinity. The further restrictions imposed by the first postulate are that both ψ and its first derivative with respect to any particular co-ordinate, $\partial\psi/\partial q_i$, should be continuous. It is not at once obvious why this should be so, but experience has shown that they are necessary conditions.

The interpretation of ψ as embodied in the postulate can be simplified by considering a single particle in a one-dimensional box where only one co-ordinate is involved. The appropriate part of the postulate could be taken to read, in this context, 'ψ is interpreted physically by $\psi\psi^* dx$ being the probability of finding the particle with co-ordinates lying between x and $x + dx$.' That is to say, that it

PARTICLE IN A BOX 41

is the probability of finding the particle in the distance element dx.
Finally, the last sentence of the postulate is simply a statement of the normalisation condition.

It has already been pointed out in the previous section that physically observable properties such as momentum are associated with operators and this situation is embodied in the second postulate of quantum mechanics.

Postulate 2

To every observable physical property of a system there corresponds a linear Hermitian operator and the physical properties of the observable may be inferred from the mathematical properties of the corresponding operator.
If the observable is a sharp quantity its value is given by

$$G\psi = g\psi \quad (3.17)$$

where G is an operator and g is a real constant which is the value of the observable.

If the observable is an unsharp quantity, its average value, $\bar{g}$, is given by

$$\bar{g} = \frac{\int \psi^* G\psi \, d\tau}{\int \psi^* \psi \, d\tau} \quad (3.18)$$

This postulate contains some new ideas which can probably best be clarified by considering some examples in depth. The first thing to establish is the method of construction of quantum mechanical operators.

The classical expression for the observable of interest is first written in terms of co-ordinates, momenta and time. All the co-ordinates and time (and hence functions which are dependent only on co-ordinates and time) are left alone and anywhere that a momentum appears in the expression it is replaced by the operator $(h/2\pi i) \cdot \partial/\partial q$.

Suppose, for example, that the quantum mechanical operator corresponding to total energy is required. Classically, the total energy, E, is equal to the sum of the kinetic energy, T, and the

potential energy, V. The classical expression for E is thus
$$E = T + V \tag{3.19}$$
The potential energy of the particle depends only on its position therefore V is a function only of the co-ordinates. The kinetic energy is given by
$$T = \tfrac{1}{2}mv^2 = \frac{(mv)^2}{2m} = \frac{p^2}{2m}$$
Equation 3.19 may be written in terms of co-ordinates and momenta as
$$E = \frac{p^2}{2m} + V \tag{3.20}$$
To construct the total energy operator, terms which are only functions of the co-ordinates (V in this case) are left unaltered and momentum terms have to be replaced by the momentum operator. As $p^2 = p \cdot p$,

$p \cdot p$ is replaced by
$$\frac{h}{2\pi i}\frac{\partial}{\partial q}\left(\frac{h}{2\pi i}\frac{\partial}{\partial q}\right)$$

In this case the operator operates on itself. Thus $\partial/\partial q$ of $(h/2\pi i).\partial/\partial q$ is
$$\frac{h}{2\pi i} \cdot \frac{\partial^2}{\partial q^2}$$
and multiplying this result by the remaining factor of $h/2\pi i$ gives
$$-\frac{h^2}{4\pi^2}\frac{\partial^2}{\partial q^2}$$
as $i^2 = -1$.

The total energy operator is obtained by substituting this term for p^2 in equation 3.20 which gives the operator as
$$\left[-\frac{h^2}{8\pi^2 m}\frac{\partial^2}{\partial q^2} + V\right]$$
This particular operator for total energy is called the *Hamiltonian operator* and is usually represented by the symbol H.

If interest lies in the total energy of the system, the second postulate states that if
$$H\psi = E\psi \tag{3.21}$$
where E is a real number then that value of E is the energy of the system when it is in the state represented by ψ.

PARTICLE IN A BOX

Before considering an application of this concept it might be useful to emphasise that ψ cannot be cancelled out in equation 3.21 because H is an operator and not simply a multiplicative factor. A simple example should illustrate this point. Consider the operator d/dx operating on the function e^{ax}, then

$$\frac{d}{dx} e^{ax} = a\, e^{ax}$$

This relation is analogous to equation 3.21 and it would be nonsensical to cancel out the term e^{ax}.

A further point contained in the second postulate is that an equation like 3.21 will only give the value of the total energy of the system if it is a sharp quantity. A sharp quantity is one which has a definite value for a given state of the system and does not vary with the co-ordinates. An example of such a quantity is, in fact, the energy of a particle in a one-dimensional box. It has been shown already that for each particular state of the system (defined by the value of the quantum number) the energy has a definite value. It should thus be possible to apply equation 3.21 to the particle in the box in order to determine the values of the energy of the system and it will be instructive to do this.

In the one-dimensional box the particle moves in the x direction so that there is only one co-ordinate to consider. Hence the generalised term $\partial^2/\partial q^2$ becomes d^2/dx^2 and the Hamiltonian operator takes the form

$$H = \left[-\frac{h^2}{8\pi^2 m} \frac{d^2}{dx^2} + V \right]$$

For the particle in the box, however, the potential energy is zero and the Hamiltonian reduces to

$$H = -\frac{h^2}{8\pi^2 m} \frac{d^2}{dx^2}$$

Substituting for H in equation 3.21

$$-\frac{h^2}{8\pi^2 m} \frac{d^2\psi}{dx^2} = E\psi$$

or

$$\frac{h^2}{8\pi^2 m} \frac{d^2\psi}{dx^2} + E\psi = 0$$

which may be written

$$\frac{d^2\psi}{dx^2} + \frac{8\pi^2 m}{h^2} E\psi = 0$$

44 PARTICLE IN A BOX

This equation is, of course, the Schrödinger equation for the particle in a one-dimensional box (equation 3.1) which has already been solved earlier in the chapter to find suitable functions ψ_n and the corresponding energies E_n.

The equation

$$H\psi = E\psi$$

is thus just a form of the Schrödinger equation. It is known as an eigenvalue equation where the function ψ is an eigenfunction of the operator H and E is the corresponding eigenvalue.

To summarise so far, it can be said that an observable quantity has an associated quantum mechanical operator. If, when the operator acts on the wave function, the result is a real number multiplied by the original wave function, i.e., if

$$G\psi = g\psi$$

where G is an operator and g is a real number, then the observable is a sharp quantity with the value g. That is to say, the successive measurements of the observable for a given state of the system will always give the value g. Normally there are several different quantised values corresponding to different states of the system, these different states being described by different wave functions. Thus

$$G\psi_1 = g_1\psi_1$$
$$G\psi_2 = g_2\psi_2 \quad \text{etc.}$$

It has already been shown that for the particle in the box there are several possible states, each described by a wave function, ψ_n, and that each state has its own energy level, E_n, so that

$$H\psi_1 = E_1\psi_1$$
$$H\psi_2 = E_2\psi_2 \quad \text{etc.}$$

The idea of the eigenvalue equation is now applied in an effort to determine the momentum of the particle in the box. At this stage it is not known whether the momentum is a sharp quantity or not, but this will emerge from the result of operating on the wave function with the momentum operator.

The momentum operator is $(h/2\pi i) \cdot \partial/\partial q$ but as the particle in the box moves only in the x direction, for the present purpose the operator takes the form $(h/2\pi i) \cdot d/dx$. Now, if

$$\frac{h}{2\pi i}\frac{d}{dx}(\psi) = \text{(a real number)} \ \psi$$

then the real number is the value of the momentum of the particle.

PARTICLE IN A BOX

For the particle in the box

$$\psi_n = \sqrt{\left(\frac{2}{a}\right)} \sin \frac{n\pi}{a} x$$

Operating on this function with the momentum operator

$$\frac{h}{2\pi i} \frac{d}{dx} \left[\sqrt{\left(\frac{2}{a}\right)} \sin \frac{n\pi}{a} x \right] = \frac{h}{2\pi i} \frac{n\pi}{a} \sqrt{\left(\frac{2}{a}\right)} \cos \frac{n\pi}{a} x$$

It can readily be seen that this result is not a real number multiplied by ψ_n and it is therefore not possible to determine the momentum of the particle in this way. Furthermore, the momentum cannot be a sharp quantity. It is instructive, however, to try to determine the square of the momentum. The operator to use is

$$\frac{h}{2\pi i} \frac{d}{dx} \left(\frac{h}{2\pi i} \frac{d}{dx} \right) = -\frac{h^2}{4\pi^2} \frac{d^2}{dx^2}$$

Operating on ψ_n,

$$-\frac{h^2}{4\pi^2} \frac{d^2}{dx^2} \left[\sqrt{\left(\frac{2}{a}\right)} \sin \frac{n\pi}{a} x \right] = -\frac{h^2}{4\pi^2} \frac{n\pi}{a} \sqrt{\left(\frac{2}{a}\right)} \frac{d}{dx} \left(\cos \frac{n\pi}{a} x \right)$$

$$= -\frac{h^2}{4\pi^2} \frac{n^2 \pi^2}{a^2} \sqrt{\left(\frac{2}{a}\right)} \left(-\sin \frac{n\pi}{a} x \right)$$

$$= \frac{n^2 h^2}{4a^2} \psi_n$$

This operation has given a real number (not imaginary as i is not involved) multiplied by ψ_n and the real number is therefore equal to the square of the momentum which must be a sharp quantity. Thus

$$p^2 = n^2 h^2 / 4a^2 \tag{3.22}$$

Notice that as n is a quantum number, p^2 is quantised in units of $h^2/4a^2$.

Although the value of p could not be obtained from an eigenvalue equation as it is not a sharp quantity, its average value may be obtained from equation 3.18 which was given in the second postulate.

$$\bar{g} = \frac{\int \psi^* G \psi \, d\tau}{\int \psi^* \psi \, d\tau} \tag{3.18}$$

Applying this equation in the appropriate form to determine the

46 PARTICLE IN A BOX

average value of the momentum of the particle in the box

$$\bar{p} = \frac{\int_0^a \psi^* \frac{h}{2\pi i} \frac{d}{dx} \psi \, dx}{\int_0^a \psi^* \psi \, dx} \qquad (3.23)$$

As the wave function given by equation 3.10 does not contain any complex quantities, the conjugate of the function is the same as the function itself and equation 3.23 may be written

$$\bar{p} = \frac{\int_0^a \psi \frac{h}{2\pi i} \frac{d}{dx} \psi \, dx}{\int_0^a \psi^2 \, dx} \qquad (3.24)$$

If the normalised wave function is used

$$\int_0^a \psi^2 \, dx = 1$$

and equation 3.24 becomes

$$\bar{p} = \int_0^a \left\{ \left[\sqrt{\left(\frac{2}{a}\right)} \sin \frac{n\pi}{a} x \right] \frac{h}{2\pi i} \frac{d}{dx} \left[\sqrt{\left(\frac{2}{a}\right)} \sin \frac{n\pi}{a} x \right] \right\} dx$$

Taking the multiplicative factors $\sqrt{(2/a)}$ and $h/2\pi i$ outside the integration sign

$$\bar{p} = \frac{2}{a} \cdot \frac{h}{2\pi i} \int_0^a \left[\left(\sin \frac{n\pi}{a} x \right) \frac{d}{dx} \left(\sin \frac{n\pi}{a} x \right) \right] dx$$

The differentiation operation inside the squared bracket is performed first, giving

$$\bar{p} = \frac{h}{a\pi i} \int_0^a \left[\left(\sin \frac{n\pi}{a} x \right) \frac{n\pi}{a} \left(\cos \frac{n\pi}{a} x \right) \right] dx$$

or

$$\bar{p} = \frac{h}{a\pi i} \cdot \frac{n\pi}{a} \int_0^a \left[\left(\sin \frac{n\pi}{a} x \right) \left(\cos \frac{n\pi}{a} x \right) \right] dx \qquad (3.25)$$

Since $\sin \alpha \cos \alpha = \frac{1}{2} \sin 2\alpha$ equation 3.25 reduces to

$$\bar{p} = \frac{nh}{2a^2 i} \int_0^a \left(\sin \frac{2n\pi}{a} x \right) dx$$

whence

$$\bar{p} = -\frac{a}{2n\pi} \cdot \frac{nh}{2a^2 i} \left[\cos \frac{2n\pi}{a} x \right]_0^a$$

PARTICLE IN A BOX 47

Remembering that n is an integer it may be seen that, when the limits of integration are inserted,

$$\bar{p} = 0$$

Accordingly, the average of a large number of measurements of p is zero. This may be understood by considering that the square of the momentum was given by equation 3.22 as

$$p^2 = n^2 h^2/4a^2 \tag{3.22}$$

whence

$$p = \pm nh/2a$$

Thus with a large number of measurements, half the observations would yield the value $p = +nh/2a$ and the other half would yield $p = -nh/2a$. It is never known in advance whether the result of a single measurement of p will give a value of $+nh/2a$ or $-nh/2a$. The particle is confined to move backwards and forwards along the x-axis. When it is travelling in the positive x direction its velocity and hence its momentum will be positive quantities. Conversely, when it is travelling in the negative x direction its velocity and momentum will be negative quantities. The result of a single measurement of momentum will thus depend on the direction of travel of the particle at the time when the measurement is made. As a result there is an uncertainty in the knowledge of the momentum of the particle which is equal to the interval between the two values. Denoting this uncertainty as Δp,

$$\Delta p = nh/a$$

There is also an uncertainty in the position of the particle. It is only known that the particle is somewhere in the box, i.e., between the limits $x = 0$ and $x = a$. The uncertainty in position, Δx, is thus given by

$$\Delta x = a$$

The product of the two uncertainties gives

$$\Delta p . \Delta x = (nh/a) . a = nh$$

The smallest value of this product will occur when $n = 1$, giving

$$\Delta p . \Delta x \approx h$$

which is a statement of the Heisenberg Uncertainty Principle.

The remaining concept in the second postulate which still requires explanation is the meaning of the term Hermitian as applied to quantum mechanical operators. It is required that the average values calculated from equation 3.18 be real numbers. With normalised

48 PARTICLE IN A BOX

wave functions, equation 3.18 may be written

$$\bar{g} = \int \psi^* G \psi \, d\tau \qquad (3.26)$$

Taking complex conjugates

$$\bar{g}^* = \int \psi G^* \psi^* \, d\tau \qquad (3.27)$$

If, however, $\bar{g}$ is a real number, then $\bar{g} = \bar{g}^*$ and from equations 3.26 and 3.27

$$\int \psi^* G \psi \, d\tau = \int \psi G^* \psi^* \, d\tau$$

More generally it can be shown that the operator must satisfy the condition

$$\int \psi_1^* G \psi_2 \, d\tau = \int \psi_1 G^* \psi_2^* \, d\tau \qquad (3.28)$$

for equation 3.18 to yield real numbers for the average values. An operator which satisfies the condition given in equation 3.28 is said to be Hermitian.

When quantum mechanical operators are being constructed it is necessary to ensure that the operator is Hermitian. For example, if the classical expression were $x \cdot p$ then the straightforward substitution of $(h/2\pi i) \cdot d/dx$ for the momentum term would give the operator

$$x \cdot \frac{h}{2\pi i} \frac{d}{dx}$$

which, if tested by equation 3.28, would be found not to be Hermitian. If, however, the classical expression were written

$$\tfrac{1}{2}(xp + px)$$

the resultant operator would be

$$\frac{1}{2}\left[x \frac{h}{2\pi i}\left(\frac{d}{dx}\right) + \frac{h}{2\pi i}\left(\frac{d}{dx}\right) x \right]$$

or

$$\frac{h}{4\pi i}\left[x\left(\frac{d}{dx}\right) + \left(\frac{d}{dx}\right) x \right]$$

and this operator is Hermitian.

PARTICLE IN A BOX 49

PARTICLE IN A RECTANGULAR THREE-DIMENSIONAL BOX

It is now useful to consider a further hypothetical problem which will introduce another technique and concept in quantum mechanics. Suppose a particle of mass m, is confined to a three-dimensional box of sides a_x, a_y and a_z. The potential energy inside the box is zero and everywhere outside the box it is infinite. The particle must therefore always be inside the box. The three-dimensional Schrödinger equation was given in Chapter 2, equation 2.71 as

$$\nabla^2 \psi + \frac{8\pi^2 m}{h^2}(E-V)\psi = 0$$

For this particular problem where $V = 0$ inside the box, the equation takes the form

$$\nabla^2 \psi + \frac{8\pi^2 m}{h^2} E\psi = 0$$

or

$$\frac{\partial^2 \psi}{\partial x^2} + \frac{\partial^2 \psi}{\partial y^2} + \frac{\partial^2 \psi}{\partial z^2} + \frac{8\pi^2 m}{h^2} E\psi = 0 \qquad (3.29)$$

In this equation ψ is a function of three variables, x, y and z. One way of attempting to solve the equation is to see if it is possible to write ψ as the product of three functions, each of which depends only on one of the variables. It is assumed, therefore, in the first instance, that

$$\psi(x, y, z) = X(x) \cdot Y(y) \cdot Z(z) \qquad (3.30)$$

which means that ψ (a function of x, y and z) is equal to the product of three functions X, Y and Z where X is a function only of x, Y is a function only of y and Z is a function only of z. Equation 3.30 may be written more simply as

$$\psi = XYZ \qquad (3.31)$$

in which the variables are not indicated. Since the functions Y and Z are independent of x, the differentiation of equation 3.31 with respect to x gives

$$\frac{\partial \psi}{\partial x} = YZ \frac{dX}{dx}$$

and further differentiation with respect to x yields

$$\frac{\partial^2 \psi}{\partial x^2} = YZ \frac{d^2 X}{dx^2} \qquad (3.32)$$

50   PARTICLE IN A BOX

Similarly,
$$\frac{\partial^2 \psi}{\partial y^2} = XZ \frac{d^2 Y}{dy^2} \tag{3.33}$$

and
$$\frac{\partial^2 \psi}{\partial z^2} = XY \frac{d^2 Z}{dz^2} \tag{3.34}$$

Substituting from equations 3.31, 3.32, 3.33 and 3.34 into equation 3.29,

$$YZ \frac{d^2 X}{dx^2} + XZ \frac{d^2 Y}{dy^2} + XY \frac{d^2 Z}{dz^2} + \frac{8\pi^2 m}{h^2} E \; XYZ = 0$$

Dividing throughout by $(8\pi^2 m/h^2) XYZ$

$$\frac{h^2}{8\pi^2 m}\left[\frac{1}{X}\frac{d^2 X}{dx^2} + \frac{1}{Y}\frac{d^2 Y}{dy^2} + \frac{1}{Z}\frac{d^2 Z}{dz^2}\right] + E = 0 \tag{3.35}$$

Consider now the case where the motion of the particle in the box is parallel to the x-axis. In this event the y and z co-ordinates will have constant values and only the x co-ordinate will vary. Under these conditions the second and third terms in the bracket of equation 3.35 will remain constant. Further, the total energy, E, of the system is a constant so that the last three terms of equation 3.35 are all constant. This being the case, the first term in x must also have a constant value. Denoting this constant value by $-E_x$,

$$\frac{h^2}{8\pi^2 m}\frac{1}{X}\frac{d^2 X}{dx^2} = -E_x \tag{3.36}$$

Similarly, let
$$\frac{h^2}{8\pi^2 m}\frac{1}{Y}\frac{d^2 Y}{dy^2} = -E_y \tag{3.37}$$

and
$$\frac{h^2}{8\pi^2 m}\frac{1}{Z}\frac{d^2 Z}{dz^2} = -E_z \tag{3.38}$$

By comparing equation 3.35 with equations 3.36, 3.37 and 3.38 it may be seen that

$$E_x + E_y + E_z = E$$

The original assumption that ψ was a product of three separate functions, each dependent on only one variable, has been justified because three separate equations, 3.36, 3.37 and 3.38, each containing only one variable have been obtained. This process is called separation of the variables.

Equation 3.36 may be written

$$\frac{h^2}{8\pi^2 m} \cdot \frac{d^2 X}{dx^2} = -E_x X$$

or
$$\frac{d^2X}{dx^2} + \frac{8\pi^2 m}{h^2} E_x X = 0 \quad (3.39)$$

This equation is of the same form as the wave equation for the one-dimensional box, equation 3.1, except that the function X occurs rather than ψ, and the energy term is E_x rather than E. The solution of equation 3.39 will thus be of the same form as the solution to equation 3.1, the normalised form of which was given by equation 3.10. Thus,

$$X = \sqrt{\left(\frac{2}{a_x}\right)} \sin \frac{n_x \pi}{a_x} x$$

Notice that the length of the box involved in this solution is the length of the side of the box in the x direction. Furthermore, the quantum number is denoted n_x as it is a quantum number appropriate to the function X. The value of the term E_x is given by an expression similar to equation 3.7, viz.,

$$E_x = \frac{n_x^2 h^2}{8m a_x^2}$$

The solutions of equations 3.37 and 3.38 together with the expressions for E_y and E_z are similarly given by

$$Y = \sqrt{\left(\frac{2}{a_y}\right)} \sin \frac{n_y \pi}{a_y} y, \quad E_y = \frac{n_y^2 h^2}{8m a_y^2}$$

$$Z = \sqrt{\left(\frac{2}{a_z}\right)} \sin \frac{n_z \pi}{a_z} z, \quad E_z = \frac{n_z^2 h^2}{8m a_z^2}$$

Remembering that
$$\psi = XYZ \quad (3.31)$$

$$\psi = \sqrt{\left(\frac{8}{v}\right)} \sin\left(\frac{n_x \pi}{a_x} x\right) \sin\left(\frac{n_y \pi}{a_y} y\right) \sin\left(\frac{n_z \pi}{a_z} z\right)$$

where
$$v = a_x \cdot a_y \cdot a_z$$

Also recalling that
$$E = E_x + E_y + E_z$$

the total energy of the particle will be given by

$$\boxed{E = \frac{h^2}{8m}\left[\frac{n_x^2}{a_x^2} + \frac{n_y^2}{a_y^2} + \frac{n_z^2}{a_z^2}\right]} \quad (3.40)$$

PARTICLE IN A BOX

There are thus three quantum numbers required to specify the energy of the particle.

In the particular case where the box is cubic,

$$a_x = a_y = a_z$$

and equation 3.40 takes the form

$$E = \frac{h^2}{8ma^2}(n_x^2 + n_y^2 + n_z^2) \qquad (3.41)$$

where a is the common value of a_x, a_y and a_z. The energy of the system now depends on the sum of the squares of the three quantum

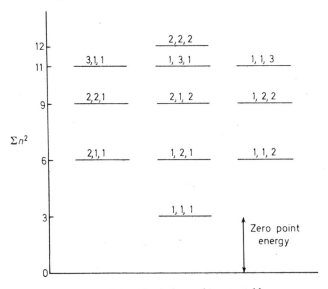

Figure 3.4 *Energy levels for a cubic potential box*

numbers and it is possible for different states of the system, as specified by particular values of the quantum numbers, to have the same energy. For example, one state of the system would be specified by $n_x = 2$, $n_y = 1$, $n_z = 1$. The state specified by $n_x = 1$, $n_y = 2$, $n_z = 1$ would be a different state because the individual values of the quantum numbers are different, but the values of

$$\Sigma n^2 = (n_x^2 + n_y^2 + n_z^2)$$

for both states are the same therefore both states have the same energy. The first few energy levels for a particle in a cubic box are shown in the energy level diagram in *Figure 3.4* where the values of the individual quantum numbers, n_x, n_y and n_z are given on the energy levels.

It will be seen from the diagram with values of Σn^2 of 6, 9 and 11 there are three states at each level with the same energy. When several states have the same energy they are said to be *degenerate*. In the cubic box problem some energy levels show *threefold* or *triple degeneracy*. Some degeneracy is removed if only two sides of the box are of equal length with the third side of a different length, and all the degeneracy is removed if all three sides of the box have different lengths. Altering the length of one side of a cubic box is analogous to subjecting the system to a perturbation and this has relevance to the Zeeman effect where a magnetic field perturbs an atom, which then loses some degeneracy giving rise to additional lines in the atomic spectrum.

4

POTENTIAL ENERGY BARRIERS

In the previous chapter, regions in which the potential energy was either zero or infinite were considered. The particle in the one-dimensional box was confined to a region of zero potential energy between the limits $x = 0$ and $x = a$ because in the regions outside these limits the potential energy was considered to be infinite. The particle was thus confined by potential energy barriers of infinite height. In this chapter attention is turned to the situation which occurs when the height of a potential energy barrier is finite, but before this problem is approached some consideration is given to the directional implications of the solutions to the Schrödinger wave equation.

DIRECTIONAL IMPLICATIONS OF WAVE FUNCTIONS

The solution to the one-dimensional Schrödinger equation was given in terms of complex exponentials in equation 2.76 as

$$\psi = C\,e^{ikx} + D\,e^{-ikx} \qquad (4.1)$$

where

$$k^2 = 8\pi^2 m(E-V)/h^2$$

As $(E-V)$ is equal to the kinetic energy, $\tfrac{1}{2}mv^2$,

$$k^2 = 4\pi^2 m^2 v^2 / h^2$$

or

$$k = 2\pi mv/h$$

so that equation 4.1 may be written

$$\psi = C\exp(2\pi i mvx/h) + D\exp[2\pi i m(-v)x/h]$$

where the negative sign of the exponent in the second term has been

POTENTIAL ENERGY BARRIERS 55

associated with the velocity of the particle. This implies that the C term corresponds to the particle having a positive velocity and that the D term corresponds to the particle having a negative velocity and thus moving in the opposite direction. The C term is usually interpreted to refer to the particle when it is moving in the positive x direction, i.e., from left to right, and the D term is considered to refer to the particle moving from right to left.

The alternative trigonometric form of the wave function which is a solution to the Schrödinger wave equation was first given in Chapter 2 by equation 2.75 and it will be recalled that the trigonometric solution

$$\psi = A\sin kx + B\cos kx \qquad (4.2)$$

corresponds to the equation for a standing wave which is produced by two progressive waves travelling in opposite directions simultaneously.

Thus, whilst the two terms of equation 4.1, the complex exponential function, may be considered separately to refer to forward and backward motion of a particle, the trigonometric function, equation 4.2, must be considered to correspond to the particle moving in opposite directions simultaneously. Needless to say it is not possible to imagine this situation physically, but it must be remembered that wave equations are only a description of the behaviour of particles in a quantum mechanical sense. It follows that in equations 4.1 and 4.2 there are two different descriptions of a problem and it often happens that one of the descriptions is more suited to a particular aspect of the problem than the other. It is obviously an advantage to choose the better of the two descriptions for a particular purpose. For example, a piece of wood could be described by stating its shape and saying that it was a wedge-shaped solid object. Alternatively, it could be described as a piece of combustible material. If the problem is to keep a door open the first description is obviously more appropriate whereas the second description would be more helpful if the problem was to light a fire.

The matter of choice of description for a problem may be illustrated by referring to the momentum of a particle in a one-dimensional box. In the last chapter an attempt was made to determine the momentum of the particle by applying the momentum operator to the wave function

$$\psi_n = \sqrt{\left(\frac{2}{a}\right)} \sin\frac{n\pi}{a} x \qquad (4.3)$$

The failure of this method showed that the momentum was not a

56 POTENTIAL ENERGY BARRIERS

sharp quantity and subsequent calculation gave the momentum, p, as

$$p = \pm nh/2a$$

showing that the momentum at any instant might be positive or negative. It is now clear that the trigonometric solution above corresponds to the particle moving in both directions simultaneously and under these conditions the momentum cannot possibly be a sharp quantity.

It is possible to convert equation 4.3 into terms of complex exponentials. Remembering (Appendix 1) that,

$$\sin \alpha = \frac{1}{2i}(e^{i\alpha} - e^{-i\alpha})$$

equation 4.3 may be written

$$\psi_n = \sqrt{\left(\frac{2}{a}\right)} \cdot \frac{1}{2i}(e^{in\pi x/a} - e^{-in\pi x/a}) \qquad (4.4)$$

In this form the first term in the bracket may be considered to represent the particle moving in the positive x direction from left to right and the second term represents the particle moving in the opposite direction. If the particle is considered to be moving only in one direction, say from left to right, then its momentum should be a well defined and hence sharp quantity. The function which represents the particle moving only from left to right will be

$$\psi_n = \sqrt{\left(\frac{2}{a}\right)} \cdot \frac{1}{2i} e^{in\pi x/a}$$

considering only the first term in equation 4.4. Operating on this function with the appropriate momentum operator

$$\frac{h}{2\pi i} \cdot \frac{d}{dx}\left[\sqrt{\left(\frac{2}{a}\right)} \cdot \frac{1}{2i} e^{in\pi x/a}\right] = \frac{h}{2\pi i} \cdot \frac{1}{2i} \cdot \sqrt{\left(\frac{2}{a}\right)} \cdot \frac{in\pi}{a} e^{in\pi x/a} = \frac{nh}{2a}\psi_n$$

This operation has yielded a real number multiplied by the original wave function and hence this real number is the momentum of the particle and

$$p = \frac{nh}{2a}$$

The momentum is positive in this case as the particle was considered to be moving in the positive x direction, i.e., with positive velocity.

It is now clear that if the direction of motion of particles is of interest in a particular problem the most appropriate description of the system will be given by a wave function expressed in terms of complex exponentials.

SINGLE POTENTIAL BARRIERS

Suppose particles travel from a region where the potential energy is zero towards a region where the potential energy is finite. If attention is restricted to the case where the motion is confined to one dimension the particles may be considered to move along the x-axis. The situation may be represented as shown in *Figure 4.1* when particles approach the potential barrier, of height U and located at $x = 0$, from the left, i.e., from negative values of x. In practice, the potential energy does not change as sharply as indicated in *Figure 4.1* which represents a simplification of true situations.

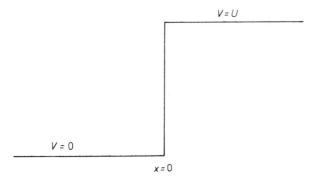

Figure 4.1 Single potential barrier of finite height

There are two regions in this problem, one corresponding to negative values of x where the potential energy is given by $V = 0$ and one corresponding to positive values of x where $V = U$. These two regions will be called region 1 and region 2 respectively.

In classical physics, particles approaching the barrier from the left would either pass over the barrier if their energy was greater than the height of the barrier, or be reflected if their energy was less than the height of the barrier. The quantum mechanical result must not be anticipated on the grounds of classical physics, but as the problem concerns the direction of motion of the particles it is apparent that the complex exponential form of the wave functions will be the more appropriate for this particular problem.

In region 1 where $V = 0$, the Schrödinger equation takes the form

$$\frac{d^2\psi_1}{dx^2} + \frac{8\pi^2 m}{h^2} E\psi_1 = 0$$

the subscript indicating that the wave function refers to region 1.

58 POTENTIAL ENERGY BARRIERS

The solution of the wave equation in exponential terms may be written

$$\psi_1 = C_1 e^{ik_1 x} + D_1 e^{-ik_1 x} \tag{4.5}$$

where

$$k_1 = 2\pi(2mE)^{\frac{1}{2}}/h \tag{4.6}$$

It should be remembered that the first term in equation 4.5 represents particles moving towards the barrier from the left, whilst the second term represents particles moving from right to left in region 1 and hence which have been reflected from the barrier. Before proceeding further it will be useful to consider each of these terms separately. Taking the first term which represents particles moving forwards and writing

$$\psi_f = C_1 e^{ik_1 x} \tag{4.7}$$

the probability of finding a particle moving forwards in an element of distance dx is given by

$$\psi_f \psi_f^* \, dx$$

where ψ_f^* is the conjugate of ψ_f. From equation 4.7

$$\psi_f \psi_f^* \, dx = C_1 e^{ik_1 x} . C_1^* e^{-ik_1 x} \, dx = C_1 C_1^* \, dx \tag{4.8}$$

where C_1^* is the conjugate of C_1.

Taking the second term of equation 4.5, the wave function for particles moving backwards may be written

$$\psi_b = D_1 e^{-ik_1 x} \tag{4.9}$$

and the probability of finding a particle moving backwards in a distance element dx is given by

$$\psi_b \psi_b^* \, dx$$

From equation 4.9

$$\psi_b \psi_b^* \, dx = D_1 e^{-ik_1 x} . D_1^* e^{ik_1 x} \, dx = D_1 D_1^* \, dx \tag{4.10}$$

The relative probability of backwards motion to forwards motion is, of course,

$$\frac{\psi_b \psi_b^* \, dx}{\psi_f \psi_f^* \, dx} = \frac{D_1 D_1^*}{C_1 C_1^*} \tag{4.11}$$

For an analysis of the barrier problem then, it may be understood that the relative values of C_1 and D_1 are important.

Turning attention to region 2 where $V = U$, the Schrödinger equation takes the form

$$\frac{d^2\psi_2}{dx^2} + \frac{8\pi^2 m}{h^2}(E-U)\psi_2 = 0$$

The solution to this equation may be written

$$\psi_2 = C_2 e^{ik_2 x} + D_2 e^{-ik_2 x} \qquad (4.12)$$

where

$$k_2 = 2\pi[2m(E-U)]^{\frac{1}{2}}/h$$

Once again the relative magnitudes of C_2 and D_2 are important. Relationships between the various C and D terms may be obtained from a consideration of the limitations which are to be imposed on wave functions if they are to be eigenfunctions of the system. It will be recalled that one condition is that the wave function must be single valued and continuous. Thus, at the point $x = 0$, ψ_1 must be equal to ψ_2 otherwise there would be two different probabilities of finding a particle at this point. Putting $\psi_1 = \psi_2$ gives, from equations 4.5 and 4.12,

$$C_1 e^{ik_1 x} + D_1 e^{-ik_1 x} = C_2 e^{ik_2 x} + D_2 e^{-ik_2 x}$$

Consideration of this relation where $x = 0$ yields

$$C_1 + D_1 = C_2 + D_2 \qquad (4.13)$$

A further condition which an eigenfunction must fulfil is that the first derivative of the function must be continuous (see Postulate 1). This means that at any point the slope of the wave function must not suddenly change. This condition may be stated as, at $x = 0$,

$$\frac{d\psi_1}{dx} = \frac{d\psi_2}{dx} \qquad (4.14)$$

Differentiating equation 4.5 with respect to x gives

$$\frac{d\psi_1}{dx} = ik_1 C_1 e^{ik_1 x} - ik_1 D_1 e^{-ik_1 x}$$

and differentiation of equation 4.12 yields

$$\frac{d\psi_2}{dx} = ik_2 C_2 e^{ik_2 x} - ik_2 D_2 e^{-ik_2 x}$$

Substituting in equation 4.14

$$ik_1(C_1 e^{ik_1 x} - D_1 e^{-ik_1 x}) = ik_2(C_2 e^{ik_2 x} - D_2 e^{-ik_2 x})$$

Applying the condition $x = 0$ and simplifying

$$k_1(C_1 - D_1) = k_2(C_2 - D_2) \qquad (4.15)$$

60 POTENTIAL ENERGY BARRIERS

From equation 4.13
$$C_1 = C_2 + D_2 - D_1 \tag{4.16}$$
and a value for D_1 may be obtained from equation 4.15 as
$$D_1 = C_1 - \frac{k_2}{k_1}C_2 + \frac{k_2}{k_1}D_2 \tag{4.17}$$
Substituting in equation 4.16
$$C_1 = C_2 + D_2 - C_1 + \frac{k_2}{k_1}C_2 - \frac{k_2}{k_1}D_2$$
or
$$2C_1 = C_2\left(1 + \frac{k_2}{k_1}\right) + D_2\left(1 - \frac{k_2}{k_1}\right)$$
whence
$$\boxed{C_1 = \frac{C_2}{2k_1}(k_1 + k_2) + \frac{D_2}{2k_1}(k_1 - k_2)} \tag{4.18}$$

An analogous expression for D_1 may be obtained by substituting equation 4.16 for C_1 in equation 4.17
$$D_1 = C_2 + D_2 - D_1 - \frac{k_2}{k_1}C_2 + \frac{k_2}{k_1}D_2$$
or
$$2D_1 = C_2\left(1 - \frac{k_2}{k_1}\right) + D_2\left(1 + \frac{k_2}{k_1}\right)$$
whence
$$\boxed{D_1 = \frac{C_2}{2k_1}(k_1 - k_2) + \frac{D_2}{2k_1}(k_1 + k_2)} \tag{4.19}$$

Having established these relationships there are two cases to consider according to whether the total energy, E, of the particles approaching the barrier is greater than or less than the height of the barrier, U. It might be useful at this stage to state the wave functions for regions 1 and 2 again so that the implications of the relative values of E and U may be more readily understood.

$$\psi_1 = C_1 e^{ik_1 x} + D_1 e^{-ik_1 x} \tag{4.5}$$
$$k_1 = 2\pi(2mE)^{\frac{1}{2}}/h \tag{4.20}$$
$$\psi_2 = C_2 e^{ik_2 x} + D_2 e^{-ik_2 x} \tag{4.12}$$
$$k_2 = 2\pi[2m(E-U)]^{\frac{1}{2}}/h \tag{4.21}$$

POTENTIAL ENERGY BARRIERS 61

Regardless of the value of U it may be seen that k_1 will always be a real quantity. In region 2, however, k_2 will be a real quantity if $E > U$, but will be an imaginary quantity if $E < U$.

Consider first, the case where $E > U$. Under these conditions k_2 is real. Since no particles approach from positive values of x, D_2 must be zero, in which case equations 4.18 and 4.19 reduce to

$$C_1 = \frac{C_2}{2k_1}(k_1 + k_2) \quad (4.22)$$

$$D_1 = \frac{C_2}{2k_1}(k_1 - k_2) \quad (4.23)$$

whence

$$\frac{D_1}{C_1} = \frac{k_1 - k_2}{k_1 + k_2}$$

Substituting from equations 4.20 and 4.21 and simplifying

$$\frac{D_1}{C_1} = \frac{E^{\frac{1}{2}} - (E - U)^{\frac{1}{2}}}{E^{\frac{1}{2}} + (E - U)^{\frac{1}{2}}} \quad (4.24)$$

Equation 4.11 gives the ratio of reflected particles to incident particles in region 1 as $D_1 D_1^*/C_1 C_1^*$. Equation 4.24 shows that D_1/C_1 is a real quantity when $E > U$ and hence

$$\frac{D_1 D_1^*}{C_1 C_1^*} = \frac{D_1^2}{C_1^2} = \frac{2E - U - 2[E(E - U)]^{\frac{1}{2}}}{2E - U + 2[E(E - U)]^{\frac{1}{2}}} \quad (4.25)$$

Dividing the numerator and denominator of equation 4.25 by E and putting $U/E = \alpha$, gives

$$\boxed{\frac{D_1 D_1^*}{C_1 C_1^*} = \frac{2 - \alpha - 2(1 - \alpha)^{\frac{1}{2}}}{2 - \alpha + 2(1 - \alpha)^{\frac{1}{2}}}} \quad (4.26)$$

The binomial expansion of $(1 - \alpha)^{\frac{1}{2}}$ gives

$$(1 - \alpha)^{\frac{1}{2}} = 1 - \frac{\alpha}{2} - \frac{\alpha^2}{8} - \frac{\alpha^3}{48} - \cdots$$

and if $\alpha \not> \frac{1}{2}$, the fourth and subsequent terms may be neglected. Substituting the first three terms of the expansion into equation 4.26 gives

$$\frac{D_1 D_1^*}{C_1 C_1^*} = \frac{\alpha^2/4}{4 - 2\alpha - \alpha^2/4} \quad (4.27)$$

If $\alpha \not> \frac{1}{2}$, then $\alpha^2/4$ is negligible with respect to $(4 - 2\alpha)$ and equation

62 POTENTIAL ENERGY BARRIERS

4.27 may be written

$$\frac{D_1 D_1^*}{C_1 C_1^*} = \frac{\alpha^2}{8(2-\alpha)} \quad (4.28)$$

Equation 4.28 shows that some particles will be reflected from the barrier provided that α is not zero. If the height of the barrier is finite α cannot be zero unless the energy of the incident particles is infinite. This means that when particles of energy E approach a potential energy barrier of height U, where $U < E$, some particles will be reflected. If $E = 2U$, then $\alpha = 0.5$ and substituting this value in equation 4.28

$$\frac{D_1 D_1^*}{C_1 C_1^*} = 0.021$$

which shows that under these conditions about 2% of the incident particles will be reflected.

Consider now the case where the energy of the incident particles is less than the height of the barrier so that $E < U$. Under these conditions k_2 is an imaginary quantity. If k_2 is imaginary, then ik_2 will be a real quantity. Putting $ik_2 = k_2'$ where k_2' is a real quantity, equation 4.12 may be written

$$\psi_2 = C_2 e^{k_2' x} + D_2 e^{-k_2' x} \quad (4.29)$$

Considering each term, it will be seen that as

$$x \to \infty$$

then
$$C_2 e^{k_2' x} \to \infty$$
and
$$D_2 e^{-k_2' x} \to 0$$

One of the conditions which an eigenfunction must fulfil is that it shall be finite everywhere. If ψ_2 is to obey this condition then C_2 must be zero, otherwise ψ_2 would be infinite when x was infinite. When $E < U$ then, ψ_2 is given by

$$\psi_2 = D_2 e^{-ik_2 x} \quad (4.30)$$

With $C_2 = 0$, equations 4.18 and 4.19 become

$$C_1 = \frac{D_2}{2k_1}(k_1 - k_2)$$

and

$$D_1 = \frac{D_2}{2k_1}(k_1 + k_2)$$

Thus,

$$\frac{D_1}{C_1} = \frac{k_1 + k_2}{k_1 - k_2}$$

POTENTIAL ENERGY BARRIERS   63

Substituting from equations 4.20 and 4.21 and simplifying

$$\frac{D_1}{C_1} = \frac{E^{\frac{1}{2}}+(E-U)^{\frac{1}{2}}}{E^{\frac{1}{2}}-(E-U)^{\frac{1}{2}}} \quad (4.31)$$

With $E < U$, $(E-U)^{\frac{1}{2}}$ is an imaginary quantity and can be written as $i(U-E)^{\frac{1}{2}}$. Equation 4.31 becomes

$$\frac{D_1}{C_1} = \frac{E^{\frac{1}{2}}+i(U-E)^{\frac{1}{2}}}{E^{\frac{1}{2}}-i(U-E)^{\frac{1}{2}}}$$

Equation 4.11 gives the ratio of reflected particles to incident particles in region 1 as $D_1 D_1^*/C_1 C_1^*$. Remembering that the conjugate of a complex quantity is obtained by replacing i with $-i$,

$$\frac{D_1 D_1^*}{C_1 C_1^*} = \frac{E^{\frac{1}{2}}+i(U-E)^{\frac{1}{2}}}{E^{\frac{1}{2}}-i(U-E)^{\frac{1}{2}}} \cdot \frac{E^{\frac{1}{2}}-i(U-E)^{\frac{1}{2}}}{E^{\frac{1}{2}}+i(U-E)^{\frac{1}{2}}} = 1$$

showing that when $E < U$ all the particles are reflected by the barrier. There is, however, a further consideration. It has already been pointed out that with $C_2 = 0$, equation 4.18 becomes

$$C_1 = \frac{D_2}{2k_1}(k_1 - k_2)$$

or

$$2k_1 C_1 = D_2(k_1 - k_2)$$

Substituting for k_1 and k_2 from equations 4.20 and 4.21 and simplifying

$$2C_1 E^{\frac{1}{2}} = D_2[E^{\frac{1}{2}} - (E-U)^{\frac{1}{2}}] \quad (4.32)$$

Once again writing $(E-U)^{\frac{1}{2}}$ as $i(U-E)^{\frac{1}{2}}$, equation 4.32 becomes

$$2C_1 E^{\frac{1}{2}} = D_2[E^{\frac{1}{2}} - i(U-E)^{\frac{1}{2}}] \quad (4.33)$$

The ratio of the probability of finding particles in region 2, inside the potential barrier, to that of finding incident particles in region 1 will be $D_2 D_2^*/C_1 C_1^*$. To have some idea of this quantity, equation 4.33 has to be multiplied on each side by the appropriate conjugates. As it is not known whether C_1 and D_2 are imaginary quantities or not, their conjugates can only be represented as C_1^* and D_2^* respectively. It will be recalled, however, that the conjugate of a number is obtained by replacing i with $-i$. In the case of the number 2, which does not contain i and is therefore not an imaginary quantity, its conjugate then is also the number 2. Thus

$$2(2^*) = 2^2 = 4$$

Similarly, for $E^{\frac{1}{2}}$,

$$E^{\frac{1}{2}}(E^{\frac{1}{2}})^* = (E^{\frac{1}{2}})^2 = E$$

64 POTENTIAL ENERGY BARRIERS

The term in the bracket of equation 4.33 is of the form $(a-ib)$ and its conjugate is therefore of the form $(a+ib)$. It has been shown in Appendix 1 that

$$(a+ib)(a-ib) = a^2 + b^2$$

so that equation 4.33 multiplied by the appropriate conjugates may be written

$$4C_1 C_1^* E = D_2 D_2^* [U + (U-E)]$$

or

$$\boxed{D_2 D_2^* = 4\frac{E}{U} C_1 C_1^*} \quad (4.34)$$

Equation 4.34 shows that when E is finite, $D_2 D_2^*$ *cannot* be zero unless U is infinite. Thus when the potential barrier has a finite height there is a probability of finding the particle *inside* the barrier.

Recapitulating the situation when $E < U$, the wave functions ψ_1 and ψ_2 for regions 1 and 2 respectively are given by equations 4.5 and 4.30 as

$$\psi_1 = C_1 e^{ik_1 x} + D_1 e^{-ik_1 x} \quad (4.5)$$

$$\psi_2 = D_2 e^{-ik_2 x} \quad (4.30)$$

At the beginning of this chapter it was pointed out that in an equation like 4.5, the C term could be interpreted as representing particles travelling from left to right and the D term as representing particles travelling from right to left. In equation 4.5, k_1 is a real quantity, being given by

$$k_1 = 2\pi(2mE)^{\frac{1}{2}}/h \quad (4.20)$$

Thus the powers to which both the exponential terms in equation 4.5 are raised are imaginary quantities, as k_1 is multiplied by i in both powers. Further, it has been shown in Chapter 2 that an equation of the form of 4.5 can be expressed in trigonometric terms as

$$\psi_1 = A \sin k_1 x + B \cos k_1 x \quad (4.35)$$

Equation 4.35 is obviously an oscillating function therefore equation 4.5 must also be an oscillating function.

In equation 4.30, however, k_2 is an imaginary quantity when $E < U$ so that ik_2 is a real quantity which may be written as k_2'. Equation 4.30 may thus be written

$$\psi_2 = D_2 e^{-k_2' x} \quad (4.36)$$

where k_2' is a real quantity. Equation 4.36 is not an oscillating function but one which decays to zero as x goes to infinity. In this case ψ_2 cannot be interpreted as an oscillating function which represents particles travelling from right to left inside the barrier but is better considered as an exponentially decaying absorption term.

THE TUNNEL EFFECT

From a consideration of the above results, where a particle with energy, E, less than the height, U, of a potential barrier may penetrate into the barrier, it seems that if the barrier is not infinitely thick the particle may have a chance of penetrating it entirely and emerging as a free particle on the other side of the barrier. Consider a potential barrier of finite width as illustrated in *Figure 4.2*. Suppose particles travel from negative values of x towards the potential barrier. The wave functions in the three regions indicated in *Figure 4.2* may be written

$$\psi_1 = C_1 e^{ik_1 x} + D_1 e^{-ik_1 x} \qquad (4.37)$$

$$\psi_2 = C_2 e^{ik_2 x} + D_2 e^{-ik_2 x} \qquad (4.38)$$

$$\psi_3 = C_3 e^{ik_3 x} + D_3 e^{-ik_3 x} \qquad (4.39)$$

where

$$k_1 = 2\pi(2mE)^{\frac{1}{2}}/h \qquad (4.40)$$

$$k_2 = 2\pi[2m(E-U_2)]^{\frac{1}{2}}/h \qquad (4.41)$$

$$k_3 = 2\pi[2m(E-U_3)]^{\frac{1}{2}}/h \qquad (4.42)$$

If attention is restricted to the case where the total energy, E, of the incident particles is somewhere between U_2 and U_3, i.e., $U_2 > E > U_3$ then k_1 and k_3 will both be real quantities. Thus ψ_1 and ψ_3 are oscillating functions in which the C terms represent particles moving from left to right and the D terms represent particles moving from right to left. If particles are incident on the barrier only from region 1 the situation is that particles approach from the left (the C_1 term). Most particles will be reflected back into region 1 (the D_1 term), but some may penetrate the barrier and continue moving to the right in region 3 (the C_3 term). As no particles approach the barrier from the right in region 3, then D_3 must be equal to zero.

The probability of finding a particle in region 3 relative to the probability of finding an incident particle in region 1 will be given by $C_3 C_3^*/C_1 C_1^*$ and this is the probability of penetration of the barrier. In order to arrive at some notion of the factors which affect this probability a relationship must be established between C_1 and C_3. This is done in a similar way to that employed in the single

66 POTENTIAL ENERGY BARRIERS

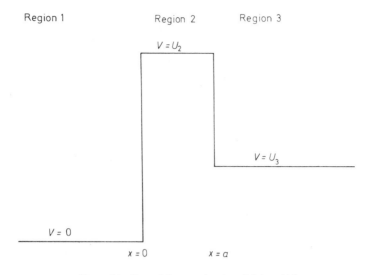

Figure 4.2   Potential energy barrier of finite width

potential barrier problem by considering the restrictions which must be imposed on ψ_1, ψ_2 and ψ_3 if these functions are to be eigenfunctions. These conditions are

$$\text{At } x = 0, \quad \psi_1 = \psi_2 \quad \text{and} \quad \frac{d\psi_1}{dx} = \frac{d\psi_2}{dx}$$

$$\text{At } x = a, \quad \psi_2 = \psi_3 \quad \text{and} \quad \frac{d\psi_2}{dx} = \frac{d\psi_3}{dx}$$

Applying these conditions to equations 4.37, 4.38 and 4.39 and remembering that $D_3 = 0$, yields:

For $\psi_1 = \psi_2$ at $x = 0$

$$C_1 + D_1 = C_2 + D_2 \tag{4.43}$$

For $d\psi_1/dx = d\psi_2/dx$ at $x = 0$

$$k_1(C_1 - D_1) = k_2(C_2 - D_2) \tag{4.44}$$

For $\psi_2 = \psi_3$ at $x = a$

$$C_2 e^{ik_2 a} + D_2 e^{-ik_2 a} = C_3 e^{ik_3 a} \tag{4.45}$$

For $d\psi_2/dx = d\psi_3/dx$ at $x = a$

$$k_2(C_2 e^{ik_2 a} - D_2 e^{-ik_2 a}) = k_3 C_3 e^{ik_3 a} \tag{4.46}$$

POTENTIAL ENERGY BARRIERS

By the same techniques as those employed in the previous section, D_1, C_2 and D_2 can be eliminated from equations 4.43, 4.44, 4.45 and 4.46 to yield a relationship between C_1 and C_3 which is (see Appendix 2),

$$C_1 = \tfrac{1}{2} C_3 \, e^{ik_3 a} \left[\left(1 + \frac{k_3}{k_1}\right) \cosh(ik_2 a) - \left(\frac{k_3}{k_2} + \frac{k_2}{k_1}\right) \sinh(ik_2 a) \right] \quad (4.47)$$

When $U_2 > E > U_3$, k_2 will be an imaginary quantity so that ik_2 will be a real quantity. Putting

$$ik_2 = k_2'$$

and substituting in equation 4.47

$$C_1 = \tfrac{1}{2} C_3 \, e^{ik_3 a} \left[\left(1 + \frac{k_3}{k_1}\right) \cosh(k_2' a) - \left(\frac{ik_3}{k_2'} + \frac{k_2'}{ik_1}\right) \sinh(k_2' a) \right]$$

or,

$$\frac{C_1}{C_3} = \tfrac{1}{2} e^{ik_3 a} \left[\left(1 + \frac{k_3}{k_1}\right) \cosh(k_2' a) - i\left(\frac{k_3}{k_2'} - \frac{k_2'}{k_1}\right) \sinh(k_2' a) \right] \quad (4.48)$$

A quantity of interest in the present context is $C_1 C_1^* / C_3 C_3^*$ and this may be obtained by multiplying equation 4.48 by its conjugate. The term in the bracket is once again of the form $(a - ib)$, and remembering that

$$(a - ib)(a + ib) = a^2 + b^2$$

$C_1 C_1^* / C_3 C_3^*$ will be given by

$$\boxed{\frac{C_1 C_1^*}{C_3 C_3^*} = \frac{1}{4}\left[\left(1 + \frac{k_3}{k_1}\right)^2 \cosh^2(k_2' a) + \left(\frac{k_3}{k_2'} - \frac{k_2'}{k_1}\right)^2 \sinh^2(k_2' a) \right]}$$

(4.49)

The probability of transmission of a particle through the barrier is, of course, $C_3 C_3^* / C_1 C_1^*$ which is the reciprocal of the quantity given by equation 4.49. The probability of transmission will not be zero unless $C_3 C_3^* / C_1 C_1^*$ is zero or, alternatively, unless $C_1 C_1^* / C_3 C_3^*$ is infinite. This will only be the case when $\cosh(k_2' a)$ and $\sinh(k_2' a)$ are infinite. Remembering that

and
$$\cosh \alpha = \tfrac{1}{2}(e^\alpha + e^{-\alpha}) \quad (4.50)$$
$$\sinh \alpha = \tfrac{1}{2}(e^\alpha - e^{-\alpha}) \quad (4.51)$$

it will be appreciated that $C_1 C_1^* / C_3 C_3^*$ will only be infinite when $k_2' a$ is infinite. Since

$$k_2' = 2\pi [2m(U_2 - E)]^{\frac{1}{2}} / h$$

the condition for zero probability of transmission through the barrier is

$$2\pi a [2m(U_2 - E)]^{1/2}/h = \infty$$

and this condition is only achieved when either $U_2 = \infty$ or $a = \infty$. Thus, unless the potential barrier is infinitely high or infinitely thick, there will always be a probability that a particle with energy less than the height of the barrier will penetrate it and this effect is called the *tunnel effect*.

To form some further idea of the phenomenon it is useful to consider the case where $k_2'a \gg 1$. Under these conditions it may be seen from equations 4.50 and 4.51 that

$$\cosh(k_2'a) \approx \sinh(k_2'a) \approx \tfrac{1}{2} e^{k_2'a}$$

In this case equation 4.49 may be written

$$\frac{C_1 C_1^*}{C_3 C_3^*} = \frac{1}{4}\left[\left(1 + \frac{k_3}{k_1}\right)^2 \cdot \frac{1}{4} e^{2k_2'a} + \left(\frac{k_3}{k_2'} - \frac{k_2'}{k_1}\right)^2 \cdot \frac{1}{4} e^{2k_2'a}\right]$$

$$= \frac{e^{2k_2'a}}{16}\left[\left(1 + \frac{k_3}{k_1}\right)^2 + \left(\frac{k_3}{k_2'} - \frac{k_2'}{k_1}\right)^2\right]$$

The probability of transmission through the barrier, $C_3 C_3^*/C_1 C_1^*$, is thus given by

$$\frac{C_3 C_3^*}{C_1 C_1^*} = \left[\frac{16}{\left(1 + \frac{k_3}{k_1}\right)^2 + \left(\frac{k_3}{k_2'} - \frac{k_2'}{k_1}\right)^2}\right] e^{-2k_2'a} \quad (4.52)$$

As the width of the barrier, a, occurs in the exponential term of equation 4.52, it may readily be understood from this equation that the probability of transmission decreases rapidly as the width of the barrier increases. The exponential factor, $\exp(-2k_2'a)$, is sometimes called the *transparency factor*.

The probability of transmission increases as the transparency factor increases, i.e., as $2k_2'a$ decreases. Now

$$2k_2'a = 4\pi a[2m(U_2 - E)]^{1/2}/h$$

so that the smaller a, m and $(U_2 - E)$, the greater is the probability of transmission. In fact, there is little chance of penetration of the energy barriers encountered on the atomic and molecular scale except for the smallest atoms and also for electrons.

For $U_2 > E > U_3$ the particles in region 1 are represented by ψ_1 and the exponential decay of ψ_2 inside the barrier leads to the emergence of an attenuated wave function (ψ_3) on the right-hand side of the barrier. The situation is analogous in some respects to the passage of light through an absorbing medium where the emergent beam is of lower intensity than the incident beam.

The tunnel effect applies in several cases in physical chemistry, the best known example being probably that of radioactive decay. Uranium-238 decays by emitting α-particles with energies of 0.67 pJ (4.2 MeV). If an α-particle approaches a nucleus there is repulsion arising from the positive charges on the nucleus and the α-particle,

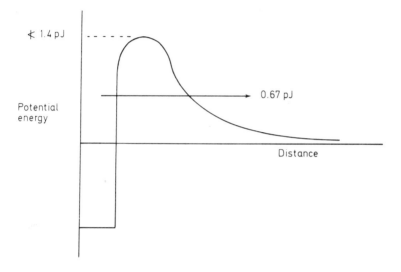

Figure 4.3 Variation of the potential energy of α-particles around a nucleus

and the force of this repulsion follows Coulomb's inverse-square law. As the α-particle gets nearer and nearer to the nucleus the potential energy rises due to this Coulomb interaction. At very short distances, however, the attractive forces binding the nucleons together overcome the repulsive force and the potential energy falls. A plot of the variation of potential energy around a nucleus is shown in *Figure 4.3* and it may be seen to approximate to the idealised situation represented in *Figure 4.2*.

Experiments have been carried out in which uranium-238 nuclei have been bombarded with 1.4 pJ (8.78 MeV) α-particles and the subsequent scattering of the α-particles due to the repulsive forces has been observed. These observations showed that the Coulomb law is obeyed at least up to energies of 1.4 pJ. The height of the energy barrier which an α-particle must pass in order to escape from the nucleus is thus at least 1.4 pJ, but the energies of the α-particles emitted in the radioactive decay of the nucleus are only 0.67 pJ. These particles can only have escaped by passing through the barrier.

The tunnel effect is also encountered in reaction kinetics when there is sometimes tunnelling through the activation energy barrier.

5

APPLICATIONS OF ONE-DIMENSIONAL MODELS

Some of the concepts which have been developed in the previous two chapters can serve as simple models for real molecular systems in certain contexts. The approximations are sometimes a little crude but they provide some insight, in a very simple fashion, to the way in which quantum mechanical concepts can give an understanding of molecular behaviour.

CONJUGATED SYSTEMS

For molecules with conjugated systems of double bonds (such as polyenes with an even number of carbon atoms) it is known that the absorption bands in the electronic spectra shift to longer wavelengths as the number of double bonds increases. In fact, long conjugated systems are often coloured indicating that the absorption band is in the visible region of the spectrum. Approximate calculations of the longest wavelength at which an absorption maximum occurs can be made using a one-dimensional box as a model for the π electrons in the conjugated system.

In hexatriene, for example, there will be six π electrons. Considering the π orbitals to be completely delocalised, the electrons will be free to move over the whole length of the molecule. This situation is analogous to a particle in a one-dimensional box where the particle is restricted to motion in one dimension within definite limits. For the π electrons in hexatriene (or any other conjugated polyene) the length of the box is not easy to define. It is usually assumed that the π electrons are free to move about half a bond length beyond each end carbon atom and that the length of the molecule is measured directly between the end carbon atoms rather than following the zigzag of the all *trans* configuration. With these

72 APPLICATIONS OF ONE-DIMENSIONAL MODELS

assumptions, and neglecting repulsion between the π electrons, the longest wavelength at which absorption occurs in the electronic spectrum may be calculated. This wavelength corresponds to the energy absorption involved in the promotion of a π electron from the highest filled level to the lowest unfilled level. For hexatriene the

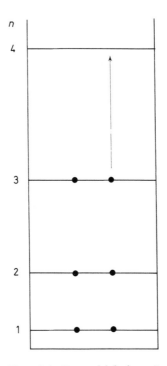

Figure 5.1 Box model for hexatriene

six π electrons are allocated to the lowest energy levels available in a one-dimensional box. Each energy level will accommodate two electrons (of opposite spin) so that the lowest three levels ($n = 1$ to $n = 3$) of the box are filled as indicated in *Figure 5.1*. The longest wavelength of absorption will thus correspond to the promotion of a π electron from the level where $n = 3$ to the level where $n = 4$. The energies of the various levels in a one-dimensional box were originally given by equation 3.7 which may be restated here as

$$E_n = \frac{n^2 h^2}{8ma^2} \tag{5.1}$$

The energy required to promote a π electron from the level $n = 3$ to

APPLICATIONS OF ONE-DIMENSIONAL MODELS 73

the level $n = 4$ is equal to the difference in energy, ΔE, between these two levels. Thus

$$\Delta E = \frac{16h^2}{8ma^2} - \frac{9h^2}{8ma^2} = \frac{7h^2}{8ma^2} \quad (5.2)$$

The frequency, v, of the absorption band is related to ΔE by (see equation 1.2)

$$v = \Delta E/h$$

whence, from equation 5.2

$$v = \frac{7h}{8ma^2} \quad (5.3)$$

The wavelength, λ, of the absorption is related to the frequency by (see equation 1.3)

$$\lambda = c/v \quad (5.4)$$

where c is the velocity of light. Substituting from equation 5.3 into equation 5.4

$$\lambda = \frac{8ma^2c}{7h} \quad (5.5)$$

For hexatriene, the distance over which the π electrons are free to move may be computed as 0.72 nm so that in equation 5.5 $a = 0.72$ nm or $a = 7.2 \times 10^{-10}$ m. The mass involved will be the mass of an electron and this, together with the other quantities in equation 5.5, is given by

$$m = 9.1 \times 10^{-31} \text{ kg}$$
$$c = 3.0 \times 10^8 \text{ m s}^{-1}$$
$$h = 6.63 \times 10^{-34} \text{ J s}$$

Expressing all these quantities in basic SI units (metres, kilogrammes, joules, seconds) and substituting in equation 5.5 should yield the result also in basic SI units. As the calculation is for a length the units of the result will thus be metres and hence

$$\lambda = \frac{8 \times 9.1 \times 10^{-31} \times (7.2)^2 \times 10^{-20} \times 3 \times 10^8}{7 \times 6.63 \times 10^{-34}} \text{ m}$$

$$= 2.44 \times 10^7 \text{ m}$$

or

$$\lambda = 244 \text{ nm}$$

The experimentally observed absorption band of hexatriene is centred at 268 nm so that in view of the simplicity of the model the

74 APPLICATIONS OF ONE-DIMENSIONAL MODELS

agreement between the calculated and observed value is quite good.

The reason for the shift to longer wavelengths of the absorption bands of conjugated polyenes as the number of double bonds increases may be seen from equation 5.5. As the length of the molecule increases so does the value of a and hence that of λ.

THE VIBRATIONAL ENERGY OF A DIATOMIC MOLECULE

When a diatomic molecule vibrates, the potential energy depends upon the position of the atoms relative to one another. The potential energy is thus related to the bond length, the relationship following a Morse curve as illustrated in *Figure 5.2*.

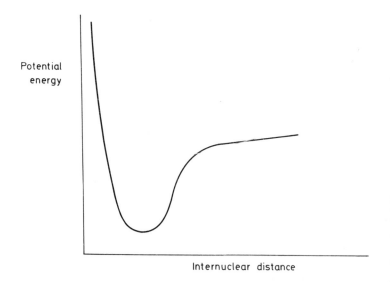

Figure 5.2   *Morse curve for a diatomic molecule*

Chemical bonds are very resistant to compression, as, shown by the incompressibility of solids, and the potential energy rises very steeply as the internuclear distance decreases. As the internuclear distance increases the bond eventually breaks, the molecule dissociates and the potential energy tends to a constant value which is that of the two isolated atoms.

The Morse potential function of *Figure 5.2* can be approximated to a simple potential function such as that illustrated in *Figure 5.3*

APPLICATIONS OF ONE-DIMENSIONAL MODELS   75

where the potential energy has a constant value of zero between $x = 0$ and $x = a$. At negative values of x the potential energy is infinite and at values of $x > a$ the potential energy has a constant finite value of U. There are thus two regions, one where the potential

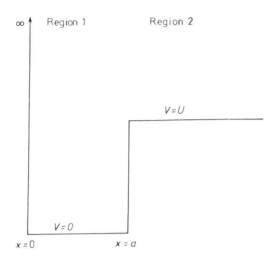

Figure 5.3 *Potential well approximation to a Morse curve*

energy is zero and one where the potential energy is finite. Denoting the wave functions for these regions as ψ_1 and ψ_2 respectively, then

$$\psi_1 = C_1 e^{ik_1 x} + D_1 e^{-ik_1 x} \tag{5.6}$$

where

$$k_1 = 2\pi(2mE)^{\frac{1}{2}}/h \tag{5.7}$$

Writing the exponential terms as trigonometric functions

$$\psi_1 = C_1(\cos k_1 x + i \sin k_1 x) + D_1(\cos k_1 x - i \sin k_1 x)$$

whence

$$\psi_1 = i(C_1 - D_1) \sin k_1 x + (C_1 + D_1) \cos k_1 x \tag{5.8}$$

Because the potential energy rises to infinity at $x = 0$, the wave function must be zero at this point. Thus at $x = 0$, $\psi_1 = 0$. Substituting these values in equation 5.8 leads to the conclusion that $(C_1 + D_1) = 0$ and making the substitution that $D_1 = -C_1$ in equation 5.8 gives

$$\psi_1 = 2iC_1 \sin k_1 x \tag{5.9}$$

APPLICATIONS OF ONE-DIMENSIONAL MODELS

The wave function ψ_2 may be written

where
$$\psi_2 = C_2 e^{ik_2 x} + D_2 e^{-ik_2 x} \tag{5.10}$$
$$k_2 = 2\pi[2m(E-U)]^{\frac{1}{2}}/h \tag{5.11}$$

In this case k_2 may be a real or imaginary quantity, depending on whether $E > U$ or $E < U$ respectively.

Consider first the case where $E < U$. Under these conditions k_2 will be an imaginary quantity but ik_2 will be a real quantity. Putting $ik_2 = k_2'$ where

$$k_2' = 2\pi[2m(U-E)]^{\frac{1}{2}}/h \tag{5.12}$$

equation 5.10 may be written in the form

$$\psi_2 = C_2 e^{k_2' x} + D_2 e^{-k_2' x} \tag{5.13}$$

Considering each term in equation 5.13 it will be seen that as

$$x \to \infty \qquad C_2 e^{k_2' x} \to \infty \quad \text{and} \quad D_2 e^{-k_2' x} \to 0$$

An eigenfunction must be finite for any value of x, so that for ψ_2 to obey this condition C_2 must be zero otherwise ψ_2 would be infinite when x was infinite. For $E < U$ then ψ_2 is given by

$$\psi_2 = D_2 e^{-k_2' x} \tag{5.14}$$

At $x = a$, $\psi_1 = \psi_2$ and applying this condition to equations 5.9 and 5.14

$$2iC_1 \sin k_1 a = D_2 e^{-k_2' a} \tag{5.15}$$

Furthermore, at $x = a$, $d\psi_1/dx = d\psi_2/dx$ which condition leads to

$$2iC_1 k_1 \cos k_1 a = -D_2 k_2' e^{-k_2' a} \tag{5.16}$$

Dividing equation 5.15 by equation 5.16

or
$$\frac{1}{k_1} \tan k_1 a = -\frac{1}{k_2'}$$

$$\tan k_1 a = -\frac{k_1}{k_2'} \tag{5.17}$$

Substituting for k_1 and k_2' from equations 5.7 and 5.12

or
$$\tan\left[\frac{2\pi a}{h}(2mE)^{\frac{1}{2}}\right] = -\frac{2\pi(2mE)^{\frac{1}{2}}}{h} \cdot \frac{h}{2\pi[2m(U-E)]^{\frac{1}{2}}}$$

$$\tan\left[\frac{2\pi a}{h}(2mE)^{\frac{1}{2}}\right] = -\frac{E^{\frac{1}{2}}}{(U-E)^{\frac{1}{2}}} \tag{5.18}$$

The values of E which satisfy equation 5.18 may be derived

APPLICATIONS OF ONE-DIMENSIONAL MODELS 77

graphically by plotting the left-hand side of the equation against E and also plotting the right-hand side of the equation against E. The two plots will then cross at a value of E which satisfies equation 5.18. The left-hand side of equation 5.18 will give a tangent graph, but as E increases, the spacing of the tangent plots will increase because they are the tangents of a function which includes the *square root* of E. The right-hand side of equation 5.18 will give a curve which has the value zero when $E = 0$ and tends to minus infinity as $E \to U$. The graphs for $U = 5 \times 10^{-20}$ J, $a = 10^{-10}$ m and $m = 5 \times 10^{-27}$ kg are illustrated in *Figure 5.4*.

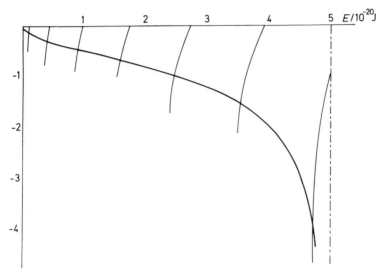

Figure 5.4 *Graphical solution of equation 5.18*

The only values of E allowed are those corresponding to the points of intersection of the curves in *Figure 5.4*. It will thus be appreciated that when $E < U$ the vibrational energy of the system is quantized.

It is interesting at this point to note the implication of equation 5.18 for the case where the potential energy at $x = a$ rises to infinity. Under these conditions

$$U = \infty \quad \text{and} \quad \frac{E^{\frac{1}{2}}}{(U-E)^{\frac{1}{2}}} = 0$$

Thus the left-hand side of equation 5.18 must also be equal to zero.

$$\tan\left[\frac{2\pi a}{h}(2mE)^{\frac{1}{2}}\right] = 0 \qquad (5.19)$$

78 APPLICATIONS OF ONE-DIMENSIONAL MODELS

The tangent of an angle is zero when the angle is an integral multiple of π radians so that the condition for equation 5.19 to be true is

$$\frac{2\pi a}{h}(2mE)^{\frac{1}{2}} = n\pi \tag{5.20}$$

where n is an integer.

Equation 5.20 leads to

$$\frac{4a^2}{h^2} \cdot 2mE = n^2$$

or

$$E = \frac{n^2 h^2}{8ma^2}$$

which is the expression giving the energies of a particle confined to a one-dimensional box with walls of infinite height (see equation 5.1).

It is now necessary to consider the situation when $E > U$ for the problem depicted by *Figure 5.3*.

The wave function ψ_2 is given by equation 5.10 as

$$\psi_2 = C_2 e^{ik_2 x} + D_2 e^{-ik_2 x} \tag{5.10}$$

or, in terms of trigonometric functions,

$$\psi_2 = i(C_2 - D_2)\sin k_2 x + (C_2 + D_2)\cos k_2 x \tag{5.21}$$

By choosing suitable values of some constant, R, and some angle, α, the coefficients of equation 5.21 may be expressed by

$$i(C_2 - D_2) = R\cos\alpha \quad \text{and} \quad (C_2 + D_2) = R\sin\alpha$$

when equation 5.21 may be written in the form

$$\psi_2 = R\sin k_2 x \cos\alpha + R\cos k_2 x \sin\alpha$$

or

$$\psi_2 = R\sin(k_2 x + \alpha) \tag{5.22}$$

The wave function, ψ_1, is given by equation 5.9 as

$$\psi_1 = 2iC_1 \sin k_1 x \tag{5.9}$$

and applying the continuity conditions at $x = a$ gives

$$\psi_1 = \psi_2 \quad \text{and} \quad d\psi_1/dx = d\psi_2/dx$$

Thus, from equations 5.9 and 5.22

$$2iC_1 \sin k_1 a = R\sin(k_2 a + \alpha) \tag{5.23}$$
$$2iC_1 k_1 \cos k_1 a = Rk_2 \cos(k_2 a + \alpha) \tag{5.24}$$

APPLICATIONS OF ONE-DIMENSIONAL MODELS 79

Dividing equation 5.23 by equation 5.24 yields

$$\frac{1}{k_1} \tan k_1 a = \frac{1}{k_2} \tan(k_2 a + \alpha) \tag{5.25}$$

and this equation may be satisfied for any values of k_1 and k_2 by making an appropriate choice of α. It will be remembered that α is related to C_2 and D_2 which are arbitrary constants that can have any value and so α can have any value. Thus for *any* given values of k_1 and k_2 it is possible to adjust the value of α so that equation 5.25 is obeyed.

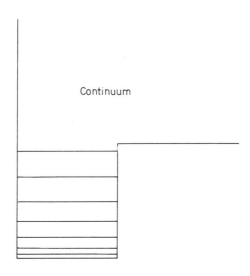

Figure 5.5   Allowed energy levels

As k_1 and k_2 involve E, the energy of the system, the above result implies that E can have any value. When $E > U$, therefore, E can have any value at all and the energy of the system is not quantized.

The overall situation relating to *Figure 5.3* is that when $E < U$ the energy is quantized and when $E > U$ the energy can have a continuous range of values, as illustrated in *Figure 5.5* for the values of U, a and m given previously.

Considering *Figure 5.5* as an approximation to the Morse curve for a diatomic molecule (*Figure 5.2*), it would be anticipated that the vibrational energies of a diatomic molecule, which are less than the dissociation energy, would be quantized.

APPLICATIONS OF ONE-DIMENSIONAL MODELS

THE ELECTRONIC ENERGY OF A DIATOMIC MOLECULE

Before considering a diatomic molecule it will be an advantage to consider how the potential energy of an electron in an atom might be represented by a simple model. In the simplest case of a hydrogen-like atom which only has one electron, the potential energy of the electron depends on its distance from the nucleus and is equal to the work required to remove the electron from its position to infinity. The potential energy, V, of any particle is given by the expression

$$\frac{dV}{dx} = -F \qquad (5.26)$$

where F is the force acting on the particle. To obtain the value of the potential energy at any particular point, equation 5.26 must be integrated. In the particular case described above, the force acting on the electron is given by Coulomb's law and for a nucleus of charge $+Ze$ and an electron of charge $-e$,

$$F = -\frac{Ze^2}{4\pi\varepsilon_0 x^2}$$

where ε_0 is the permittivity of a vacuum and x is the distance of the electron from the nucleus. Substituting in equation 5.26 and integrating

$$V = \frac{Ze^2}{4\pi\varepsilon_0 x^2}\,dx$$

or

$$V = -\frac{Ze^2}{4\pi\varepsilon_0 x} + \text{constant}$$

The constant of integration is evaluated by establishing the condition for zero potential energy. In this case the electron is considered to have zero potential energy when it is an infinite distance away from the nucleus, i.e., $V = 0$ when $x = \infty$. The value of the integration constant is thus zero and

$$\boxed{V = -\frac{Ze^2}{4\pi\varepsilon_0 x}} \qquad (5.27)$$

Equation 5.27 shows that the variation of potential energy with distance takes the form of a hyperbola as illustrated in *Figure 5.6*.

The formation of the hydrogen molecule ion, H_2^+, may be considered as the juxtaposition of two hydrogen nuclei which share an electron. The potential functions for each nucleus will be as shown

APPLICATIONS OF ONE-DIMENSIONAL MODELS 81

in *Figure 5.6* and as the two nuclei approach each other, each affects the potential function of the other. The modifications of the potential functions with approach of the nuclei are illustrated in *Figure 5.7* where the continuous lines represent the modified potential function. The potential functions shown in *Figure 5.7(c)* may be approximated to the simpler box-like function illustrated in *Figure 5.8* where the outside walls are considered to rise to infinity for the sake of simplification.

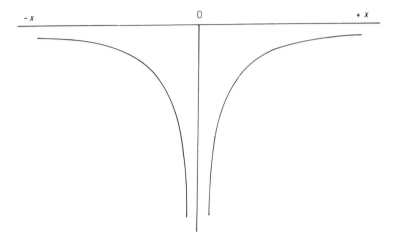

Figure 5.6 *Potential energy of an electron in a hydrogen-like atom*

There are three regions to be considered in this problem, first $-b < x < -a$ where $V = 0$, second $-a < x < +a$ where $V = U$ and third $+a < x < +b$ where $V = 0$. The wave functions for these three regions will be denoted ψ_1, ψ_2 and ψ_3 respectively. These wave functions will be given by the expressions

$$\psi_1 = C_1 e^{ik_1 x} + D_1 e^{-ik_1 x} \tag{5.28}$$

$$k_1 = 2\pi(2mE)^{\frac{1}{2}}/h \tag{5.29}$$

$$\psi_2 = C_2 e^{ik_2 x} + D_2 e^{-ik_2 x} \tag{5.30}$$

$$k_2 = 2\pi[2m(E-U)]^{\frac{1}{2}}/h \tag{5.31}$$

$$\psi_3 = C_3 e^{ik_1 x} + D_3 e^{-ik_1 x} \tag{5.32}$$

ψ_1 may be expressed in trigonometric terms as

$$\psi_1 = C_1(\cos k_1 x + i \sin k_1 x) + D_1(\cos k_1 x - i \sin k_1 x)$$

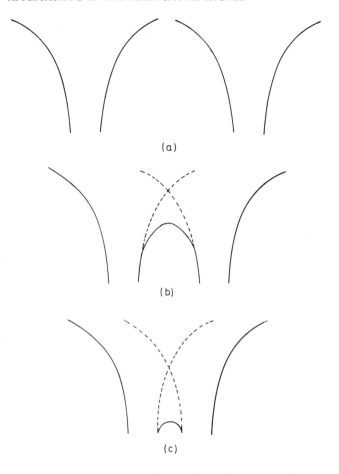

Figure 5.7 *Formation of a hydrogen molecule ion*

or
$$\psi_1 = i(C_1 - D_1)\sin k_1 x + (C_1 + D_1)\cos k_1 x \tag{5.33}$$

Quantities A_1 and α may be chosen such that

$$i(C_1 - D_1) = A_1 \cos k_1 \alpha \quad \text{and} \quad (C_1 + D_1) = A_1 \sin k_1 \alpha$$

Substituting in equation 5.33

$$\psi_1 = A_1 \sin k_1 x \cos k_1 \alpha + A_1 \cos k_1 x \sin k_1 \alpha$$

or
$$\psi_1 = A_1 \sin k_1(x + \alpha) \tag{5.34}$$

APPLICATIONS OF ONE-DIMENSIONAL MODELS

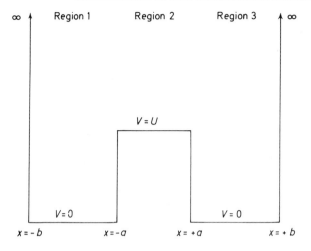

Figure 5.8 Double well model

The potential energy at $x = -b$ rises to infinity so that ψ_1 must be zero at this point. Thus the value of α must be $+b$ and equation 5.34 may be written

$$\psi_1 = A_1 \sin k_1(x+b) \tag{5.35}$$

By an exactly analogous argument which involves putting $\psi_3 = 0$ at $x = +b$, ψ_3 may be shown to be given by

$$\psi_3 = A_3 \sin k_1(x-b) \tag{5.36}$$

At the boundaries between the regions the wave functions for the regions on each side of the boundary must be equal, as must the first derivatives of the wave functions. Thus at $x = -a$, $\psi_1 = \psi_2$ and $d\psi_1/dx = d\psi_2/dx$. Applying these conditions to equations 5.35 and 5.30

$$A_1 \sin k_1(b-a) = C_2 e^{-ik_2 a} + D_2 e^{ik_2 a} \tag{5.37}$$

and

$$A_1 k_1 \cos k_1(b-a) = ik_2(C_2 e^{-ik_2 a} - D_2 e^{ik_2 a}) \tag{5.38}$$

At $x = +a$, $\psi_2 = \psi_3$ and $d\psi_2/dx = d\psi_3/dx$ and from equations 5.30 and 5.36

$$A_3 \sin k_1(a-b) = C_2 e^{ik_2 a} + D_2 e^{-ik_2 a} \tag{5.39}$$

and

$$A_3 k_1 \cos k_1(a-b) = ik_2(C_2 e^{ik_2 a} - D_2 e^{-ik_2 a}) \tag{5.40}$$

Dividing equation 5.37 by equation 5.39

$$-\frac{A_1}{A_3} = \frac{C_2 e^{-ik_2 a} + D_2 e^{ik_2 a}}{C_2 e^{ik_2 a} + D_2 e^{-ik_2 a}} \tag{5.41}$$

84 APPLICATIONS OF ONE-DIMENSIONAL MODELS

Dividing equation 5.38 by equation 5.40

$$\frac{A_1}{A_3} = \frac{C_2 e^{-ik_2 a} - D_2 e^{ik_2 a}}{C_2 e^{ik_2 a} - D_2 e^{-ik_2 a}} \quad (5.42)$$

From equations 5.41 and 5.42

$$\frac{C_2 e^{-ik_2 a} + D_2 e^{ik_2 a}}{C_2 e^{ik_2 a} + D_2 e^{-ik_2 a}} = \frac{-C_2 e^{-ik_2 a} + D_2 e^{ik_2 a}}{C_2 e^{ik_2 a} - D_2 e^{-ik_2 a}} \quad (5.43)$$

There are only two relationships between C_2 and D_2 which satisfy equation 5.43 and these are either $C_2 = D_2$ or $C_2 = -D_2$. Putting $C_2 = D_2$ in equation 5.41 leads to $A_1 = -A_3$ and putting $C_2 = -D_2$ in equation 5.41 leads to $A_1 = A_3$.

There are thus two sets of wave equations. The first set corresponds to $C_2 = D_2$ and $A_1 = -A_3$. Applying these results to equations 5.35, 5.30 and 5.36 gives

$$\boxed{\psi_1 = A_1 \sin k_1(x+b)} \quad (5.44)$$

$$\boxed{\psi_2 = C_2(e^{ik_2 x} + e^{-ik_2 x})} \quad (5.45)$$

$$\boxed{\psi_3 = -A_1 \sin k_1(x-b)} \quad (5.46)$$

This set of equations is *symmetric* about $x = 0$. This means that the value of ψ_1 for a value of x between $-a$ and $-b$ is equal to the value of ψ_3 for a value of x of the same magnitude between $+a$ and $+b$. Similarly, the value of ψ_2 for a value of x between 0 and $-a$ is the same as its value for a value of x of the same magnitude between 0 and $+a$.

The second set of wave functions corresponds to $C_2 = -D_2$ and $A_1 = A_3$. Putting these results into equations 5.35, 5.30 and 5.36 gives

$$\boxed{\psi_1 = A_1 \sin k_1(x+b)} \quad (5.47)$$

$$\boxed{\psi_2 = C_2(e^{ik_2 x} - e^{-ik_2 x})} \quad (5.48)$$

$$\boxed{\psi_3 = A_1 \sin k_1(x-b)} \quad (5.49)$$

and this set of equations is *antisymmetric* about the origin. That is to

APPLICATIONS OF ONE-DIMENSIONAL MODELS 85

say, that the values of ψ_1 and ψ_2 for negative values of x are equal to *minus* the values of ψ_3 and ψ_2 for positive values of x of the same magnitudes. From the above discussion it follows that the origin is a centre of symmetry in the problem under discussion. For a symmetric function a change of the co-ordinate from $+x$ to $-x$ leaves the value of the function unchanged whilst a similar change of co-ordinate for an antisymmetric function changes the value of the function to minus its original value. In the present problem there is only one co-ordinate involved, but with a three-dimensional problem there would be three co-ordinates, x, y and z. If, for a wave function which is a function of x, y and z, the co-ordinates are changed to $-x$, $-y$ and $-z$, which operation leaves the value of the function unchanged, it will be a symmetric wave function. Such functions which are symmetric to an inversion of the co-ordinates are usually designated with a plus sign, whilst wave functions which are antisymmetric to inversion are designated with a minus sign.

In order to obtain the energy, E, of the system it is necessary to know k_1 and k_2. In the present context it is only of interest to consider the case where $E < U$, but the symmetric and antisymmetric functions will have to be considered separately.

For the symmetric solutions consider equations 5.44 and 5.45 where a superscript, $+$, indicates the symmetric nature of the functions.

$$\psi_1^+ = A_1 \sin k_1(x+b) \qquad (5.44)$$

$$\psi_2^+ = C_2(e^{ik_2 x} + e^{-ik_2 x}) \qquad (5.45)$$

When $E < U$, k_2 will be imaginary and ik_2 will be real. Putting $ik_2 = k_2'$ where

$$k_2' = 2\pi[2m(U-E)]^{\frac{1}{2}}/h \qquad (5.50)$$

equation 5.45 may be written as

$$\psi_2^+ = C_2(e^{k_2' x} + e^{-k_2' x}) \qquad (5.51)$$

At $x = -a$, $\quad \psi_1^+ = \psi_2^+ \quad$ and $\quad d\psi_1^+/dx = d\psi_2^+/dx$

Thus, from equations 5.44 and 5.51

$$A_1 \sin k_1(b-a) = C_2(e^{-k_2' a} + e^{k_2' a}) \qquad (5.52)$$

and

$$A_1 k_1 \cos k_1(b-a) = C_2 k_2'(e^{-k_2' a} - e^{k_2' a}) \qquad (5.53)$$

Dividing equation 5.52 by equation 5.53 gives

$$\frac{1}{k_1}\tan k_1(b-a) = \frac{1}{k_2'}\frac{e^{-k_2' a}+e^{k_2' a}}{e^{-k_2' a}-e^{k_2' a}} \qquad (5.54)$$

It will be recalled from Appendix 1 that

$$\coth \alpha = \frac{e^{\alpha}+e^{-\alpha}}{e^{\alpha}-e^{-\alpha}}$$

so that equation 5.54 may be written in the form

$$\tan k_1(b-a) = -\frac{k_1}{k_2'}\coth(k_2'a) \qquad (5.55)$$

Substituting for k_1 and k_2' from equations 5.29 and 5.50

$$\tan\left[\frac{2\pi(b-a)}{h}(2mE)^{\frac{1}{2}}\right] = -\frac{E^{\frac{1}{2}}}{(U-E)^{\frac{1}{2}}}\coth\left\{\frac{2\pi a}{h}[2m(U-E)]^{\frac{1}{2}}\right\} \qquad (5.56)$$

This equation may be solved by the same method as was applied to equation 5.18 in the previous section. Each side of the equation is plotted against E and the values of E at which the plots cross satisfy equation 5.56.

For the antisymmetric solutions consider equations 5.47 and 5.48 where the superscript, $-$, is added to indicate the antisymmetric nature.

$$\psi_1^- = A_1 \sin k_1(x+b) \qquad (5.47)$$
$$\psi_2^- = C_2(e^{ik_2 x} - e^{-ik_2 x}) \qquad (5.48)$$

Once again, with $E < U$, k_2 will be imaginary and equation 5.48 may be written as

$$\psi_2^- = C_2(e^{k_2' x} - e^{-k_2' x}) \qquad (5.57)$$

where k_2' is once again given by equation 5.50.

At $x = -a$, $\quad \psi_1^- = \psi_2^- \quad$ and $\quad d\psi_1^-/dx = d\psi_2^-/dx$

Thus from equations 5.47 and 5.57

$$A_1 \sin k_1(b-a) = C_2(e^{-k_2'a} - e^{k_2'a}) \qquad (5.58)$$
and
$$A_1 k_1 \cos k_1(b-a) = C_2 k_2'(e^{-k_2'a} + e^{k_2'a}) \qquad (5.59)$$

Dividing equation 5.58 by equation 5.59

$$\frac{1}{k_1}\tan k_1(b-a) = \frac{1}{k_2'}\frac{e^{-k_2'a} - e^{k_2'a}}{e^{-k_2'a} + e^{k_2'a}} \qquad (5.60)$$

The exponential fraction on the right-hand side of this equation is seen to be the reciprocal of that in equation 5.54 and hence

$$\tan k_1(b-a) = -\frac{k_1}{k_2'}\tanh(k_2'a) \qquad (5.61)$$

Once again this equation may be solved by plotting each side against E.

Equations 5.56 and 5.61 are solved in this way in *Figure 5.9* where values of U, m, a and b have been chosen so that there is only one symmetric and one antisymmetric energy level below the value of U. This will correspond to an approximate model for the hydrogen molecule ion. The corresponding symmetric and antisymmetric wave functions are plotted in *Figure 5.10*. From this figure it should be

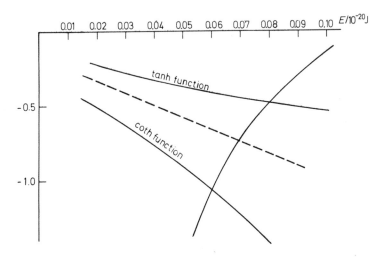

Figure 5.9   *Graphical solutions of equations* 5.56 *and* 5.61

noted that at $x = 0$, which is the point midway between the atomic nuclei, the symmetric wave function has a finite value, whilst the antisymmetric wave function is equal to zero. A zero value of the wave function means that $\psi^-\psi^{-*}$ will also be zero, which implies that there is zero probability of finding the electron between the two nuclei. In this case the two nuclei will always repel each other so that no bond will be formed. In this particular case then, the antisymmetric wave function leads to an *antibonding* situation. With the symmetric wave function there is a finite probability of finding the electron between the two nuclei which will result in a net attraction and the formation of a bond. The symmetric wave function thus leads to a *bonding* situation.

It should be further noted from *Figure 5.9* that the energy of the symmetric state, E^+, is less than that of the antisymmetric state, E^-. The significance of this point is probably better illustrated by using the model, as illustrated in *Figure 5.8*, as an approximation to a

88 APPLICATIONS OF ONE-DIMENSIONAL MODELS

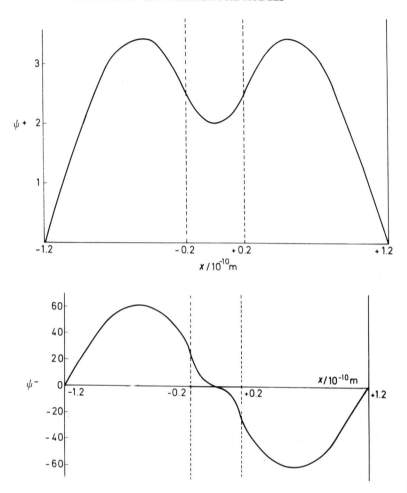

Figure 5.10 Symmetric and antisymmetric wave functions for a double well

diatomic molecule formed from multi-electron atoms. In this case the potential wells will be rather deeper and accommodate more than one energy level. *Figure 5.11* shows plots of equations 5.55 and 5.61 for $U = 5 \times 10^{-20}$ J, $m = 5 \times 10^{-27}$ kg, $a = 10^{-11}$ m and $(b-a) = 10^{-10}$ m. It can be seen from the figure that the differences between the symmetric and antisymmetric energy levels become greater as E approaches U.

APPLICATIONS OF ONE-DIMENSIONAL MODELS 89

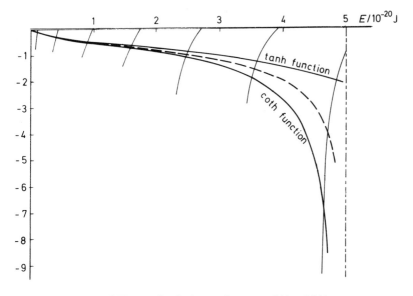

Figure 5.11 Graphical solution of equations 5.55 and 5.61

It is interesting to note the effect on the energy levels of separating the atomic nuclei to a greater extent. Greater separation of the nuclei corresponds to greater values of a in *Figure 5.8* and the effect of this on equations 5.55 and 5.61

$$\tan k_1(b-a) = -\frac{k_1}{k_2'}\coth(k_2'a) \qquad (5.55)$$

$$\tan k_1(b-a) = -\frac{k_1}{k_2'}\tanh(k_2'a) \qquad (5.61)$$

As $a \to \infty$, $\coth(k_2'a) \to \tanh(k_2'a) \to 1$

so that at large values of a, both equations 5.55 and 6.61 reduce to

$$\tan k_1(b-a) = -\frac{k_1}{k_2'} \qquad (5.62)$$

and a plot of this equation is shown as a dotted line in *Figure 5.11*. The solutions of equation 5.62 correspond to the energy levels in the atoms when they are completely separated and *Figure 5.11* shows that these levels lie between the corresponding symmetric and antisymmetric levels in the molecule. This implies that an electron in a bonding level is more stable than it would be in the isolated atom.

90 APPLICATIONS OF ONE-DIMENSIONAL MODELS

As the inner electrons in an atom will occupy the lowest energy levels *Figure 5.11* shows that the inner electrons are very little affected by combination with another atom to form a molecule. This follows from the fact that the isolated atom levels and the symmetric and antisymmetric molecular levels are very close together for the lower

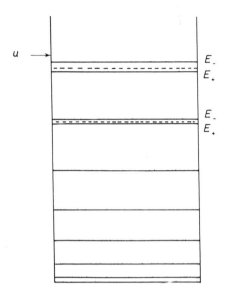

Figure 5.12   Symmetric, antisymmetric and isolated energy levels

energy levels. It is only as the value of E approaches U that significant differences occur in the isolated, symmetric and antisymmetric levels, and this is shown more clearly in *Figure 5.12*. It is the outer electrons in the highest energy levels which are stabilised by the formation of a molecule as they will be accommodated in bonding energy levels of lower energy than the corresponding level in the isolated atoms.

6

THE LINEAR HARMONIC OSCILLATOR

In the last chapter the vibrational energy of a diatomic molecule was discussed on the basis of a simplified form of potential function (*Figure 5.3*) which is a crude approximation to the Morse curve (*Figure 5.2*). Up to this point all the problems which have been discussed have involved simple step potential functions where the potential energy of the system has been considered constant over a given region of space. A better model for the vibrational energy of a diatomic molecule is the linear harmonic oscillator in which the potential energy changes continuously with distance and it is the purpose of this chapter to consider such a model.

Some of the introductory remarks in Chapter 2 concerned simple harmonic oscillation which was discussed from the classical point of view. The quantum mechanical solution of simple harmonic oscillation will now be examined.

Consider a particle of mass m executing simple harmonic motion along the x-axis, the equilibrium position of the particle being at $x = 0$. As pointed out in Chapter 2, the restoring force, F, is proportional to the displacement, but acts in the opposite direction, and hence

$$F = -kx \tag{6.1}$$

where k is the proportionality constant known as the force constant.
The potential energy of any particle is given by

$$\frac{dV}{dx} = -F \tag{6.2}$$

and hence

$$V = \int -F\,dx$$

92 THE LINEAR HARMONIC OSCILLATOR

Substituting from equation 6.1

$$V = \int kx\,dx$$

or

$$V = \tfrac{1}{2}kx^2 + \text{constant}$$

The value of the integration constant is determined by establishing the zero of potential energy. For a harmonic oscillator it is convenient to take the potential energy to be zero when the oscillator

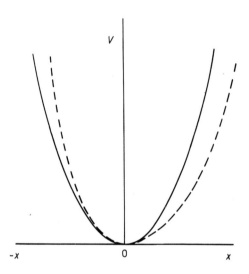

Figure 6.1 Potential energy of a linear harmonic oscillator

is at its equilibrium position so that $V = 0$ when $x = 0$. This leads to a value of zero for the integration constant and the potential energy of the oscillator is thus given by

$$V = \tfrac{1}{2}kx^2 \tag{6.3}$$

Equation 6.3 is the equation of a parabola as illustrated in *Figure 6.1* and it will be seen that, at the lower values of potential energy, this function is a better approximation to the Morse curve which is also shown in *Figure 6.1* for comparison.

To approach this problem from the quantum mechanical point of view the appropriate Schrödinger equation must be solved. It will be recalled that the Schrödinger equation for one dimension takes

the form
$$\frac{d^2\psi}{dx^2} + \frac{8\pi^2 m}{h^2}(E-V)\psi = 0 \tag{6.4}$$

Substituting for V from equation 6.3
$$\frac{d^2\psi}{dx^2} + \frac{8\pi^2 m}{h^2}(E - \tfrac{1}{2}kx^2)\psi = 0 \tag{6.5}$$

The force constant, k, in equation 6.5 may be written in an alternative form which can be derived from equations 2.5 and 2.28 in the section on simple harmonic motion in Chapter 2. These equations give

and
$$\omega = 2\pi\nu_e$$
whence
$$\omega = (k/m)^{\frac{1}{2}}$$
or
$$(k/m)^{\frac{1}{2}} = 2\pi\nu_e$$
$$k = 4\pi^2 m\nu_e^2 \tag{6.6}$$

where ν_e is the frequency of oscillation of the particle about the equilibrium position. Substituting in equation 6.5

$$\frac{d^2\psi}{dx^2} + \frac{8\pi^2 m}{h^2}(E - 2\pi^2 m\nu_e^2 x^2)\psi = 0$$

or
$$\frac{d^2\psi}{dx^2} + \left(\frac{8\pi^2 mE}{h^2} - \frac{16\pi^4 m^2 \nu_e^2}{h^2}x^2\right)\psi = 0 \tag{6.7}$$

For simplification put
$$\frac{4\pi^2 \nu_e m}{h} = \alpha \tag{6.8}$$

when equation 6.7 may be written
$$\frac{d^2\psi}{dx^2} + \left(\frac{8\pi^2 mE}{h^2} - \alpha^2 x^2\right)\psi = 0 \tag{6.9}$$

This equation may best be solved by changing the variable from x to a new variable q where
$$q = \sqrt{(\alpha)}x \tag{6.10}$$

If the variable is to be changed in this way, then in the first term of equation 6.9, ψ must be differentiated with respect to q rather than with respect to x and a relationship between d^2/dq^2 and d^2/dx^2 must be established. Now

$$\frac{d}{dx} = \frac{d}{dq} \cdot \frac{dq}{dx}$$

94 THE LINEAR HARMONIC OSCILLATOR

and from equation 6.10

$$\frac{dq}{dx} = \sqrt{\alpha}$$

Hence

$$\frac{d}{dx} = \frac{d}{dq}\sqrt{\alpha} \qquad (6.11)$$

In order to arrive at an expression for d^2/dx^2, equation 6.11 must be differentiated again with respect to x.

$$\frac{d^2}{dx^2} = \frac{d}{dx}\left(\frac{d}{dq}\sqrt{\alpha}\right)$$

Substituting from equation 6.11 for d/dx

$$\frac{d^2}{dx^2} = \frac{d}{dq}\sqrt{\alpha}\cdot\left(\frac{d}{dq}\sqrt{\alpha}\right)$$

or

$$\frac{d^2}{dx^2} = \alpha\frac{d^2}{dq^2}$$

Using this relationship and that given by equation 6.10, equation 6.9 may be written

$$\alpha\frac{d^2\psi}{dq^2} + \left(\frac{8\pi^2 mE}{h^2} - \alpha q^2\right)\psi = 0$$

or, dividing through by α

$$\frac{d^2\psi}{dq^2} + \left(\frac{8\pi^2 mE}{\alpha h^2} - q^2\right)\psi = 0 \qquad (6.11a)$$

Substituting for α from equation 6.8 into the first term in the bracket in equation 6.11a gives

$$\frac{d^2\psi}{dq^2} + \left(\frac{2E}{hv_e} - q^2\right)\psi = 0$$

which may be written

$$\frac{d^2\psi}{dq^2} + (\beta - q^2)\psi = 0 \qquad (6.12)$$

where

$$\beta = \frac{2E}{hv_e} \qquad (6.13)$$

The original problem has thus been reduced to one of solving equation 6.12. The approach is to find an approximate solution of equation 6.12 which may then be modified to give an exact solution.

THE LINEAR HARMONIC OSCILLATOR 95

APPROXIMATE SOLUTION

At very large values of q (which corresponds to very large values of x, see equation 6.10) then $q^2 \gg \beta$ and equation 6.12 could be written

$$\frac{d^2\psi}{dq^2} - q^2\psi = 0 \qquad (6.14)$$

Further, if q is very large then $(q^2 + 1) \approx q^2$ and equation 6.14 may be written

$$\frac{d^2\psi}{dq^2} - (q^2 + 1)\psi = 0 \qquad (6.15)$$

The point of these approximations is that any solution of equation 6.15 should be an approximate solution for equation 6.12 for large values of q and equation 6.15 is readily soluble. The solutions are

$$\psi = K\,e^{\frac{1}{2}q^2} \qquad (6.16)$$

and

$$\psi = K\,e^{-\frac{1}{2}q^2} \qquad (6.17)$$

where K is a constant.

In equation 6.16 it may be seen that as $q \to \infty$ then $\psi \to \infty$ so that this solution may be discarded as wave functions are required to be finite. The only solution to be considered is thus equation 6.17. This equation is an exact solution to equation 6.15 which is the form taken by equation 6.12 as $q \to \infty$. Alternatively it may be said that equation 6.17 is the asymptotic solution of equation 6.12 as $q \to \infty$.

EXACT SOLUTION

For values of $q < \infty$, equation 6.17 must be modified in some way if it is to be a solution of equation 6.12. Such modifications may sometimes be made by multiplying the result by a power series of some description. This suggests that if the constant in equation 6.17 were replaced by a power series the necessary modification would be achieved provided the power series was of the right form. Thus the solution to equation 6.12 might be something of the form

$$\psi = e^{-\frac{1}{2}q^2}.H(q) \qquad (6.18)$$

where $H(q)$ is an infinite power series given by

$$H(q) = \Sigma A_m q^m = A_0 + A_1 q + A_2 q^2 + \ldots.$$

This type of power series is known as a Hermite polynomial.

To see if equation 6.18 is a solution of equation 6.12 it may be

96 THE LINEAR HARMONIC OSCILLATOR

substituted in the second term of equation 6.12 and, after being differentiated twice, may be substituted for the first term of equation 6.12. The first task is to perform the differentiation and for simplicity, first and second derivatives will be denoted by a prime and double prime respectively. Remembering that the right-hand side of equation 6.18 is a product and must be differentiated as such, the first derivative of ψ, ψ' is given by

$$\psi' = e^{-\frac{1}{2}q^2}.H' - q e^{-\frac{1}{2}q^2}.H \tag{6.19}$$

where the Hermite polynomial is represented simply by H. This equation must be differentiated again and observing that both terms on the right-hand side are products,

or
$$\psi'' = e^{-\frac{1}{2}q^2}.H'' - 2q e^{-\frac{1}{2}q^2}.H' + q^2 e^{-\frac{1}{2}q^2}.H - e^{-\frac{1}{2}q^2}.H$$
$$\psi'' = e^{-\frac{1}{2}q^2}(H'' - 2qH' + q^2H - H) \tag{6.20}$$

Substituting from equation 6.20 and 6.18 into equation 6.12

$$e^{-\frac{1}{2}q^2}(H'' - 2qH' + q^2H - H) + (\beta - q^2) e^{-\frac{1}{2}q^2}.H = 0$$

Dividing by the exponential term

or
$$H'' - 2qH' + q^2H - H + \beta H - q^2 H = 0$$
$$H'' - 2qH' + (\beta - 1)H = 0 \tag{6.21}$$

Thus, equation 6.18 is a solution for the wave equation *provided* equation 6.21 is valid. It remains to examine the conditions under which equation 6.21 will be valid. Now, if

then
$$H = \Sigma A_m q^m \tag{6.22}$$
and
$$H' = \Sigma m A_m q^{m-1} \tag{6.23}$$
$$H'' = \Sigma m(m-1) A_m q^{m-2} \tag{6.24}$$

Substituting in equation 6.21

$$\Sigma m(m-1) A_m q^{m-2} - 2\Sigma m A_m q^m + (\beta - 1)\Sigma A_m q^m = 0 \tag{6.25}$$

Equation 6.25 has to be true for any value of q, which means that the sum of the coefficients of any particular power of q must be zero. In each of the three summations on the left-hand side of equation 6.25 a particular power of q, say q^v, will appear. The coefficient of q^v in the first term of equation 6.25 may be obtained by putting $m = v+2$ and the coefficients of q^v in the other two terms may be obtained by putting $m = v$. Performing this operation and putting the sum of these three coefficients equal to zero yields

or,
$$(v+2)(v+1)A_{(v+2)} - 2vA_v + (\beta - 1)A_v = 0$$
$$(v+2)(v+1)A_{(v+2)} = (2v+1-\beta)A_v \tag{6.26}$$

THE LINEAR HARMONIC OSCILLATOR 97

Equation 6.26 is a *recursion formula* which enables $A_{(v+2)}$ to be calculated in terms of A_v. Thus, starting with A_0, the coefficients A_2, A_4, A_6, etc., may be calculated and starting with A_1, the coefficients A_3, A_5, A_7, etc., may be calculated. In this way, two power series can be generated which are solutions for $H(q)$ in equation 6.18.

and
$$H(q) = A_0 + A_2 q^2 + A_4 q^4 + \ldots \quad (6.27)$$
$$H(q) = A_1 q + A_3 q^3 + A_5 q^5 + \ldots \quad (6.28)$$

If the two series in equations 6.27 and 6.28 were infinite series, then as $q \to \infty$ the series would diverge and their sums would become infinite. This would not be an acceptable solution as wave functions are required to be finite. The series must thus be limited in some way.

It may be seen from equation 6.26 that if $(2v+1-\beta) = 0$ for some value of v, then the coefficient $A_{(v+2)}$ will be zero as will $A_{(v+4)}$, $A_{(v+6)}$, etc. Thus when $(2v+1-\beta) = 0$, the polynomial terminates at the term $A_v q^v$ and

or
$$H(q) = A_0 + A_2 q^2 + A_4 q^4 + \ldots + A_v q^v.$$
$$H(q) = A_1 q + A_3 q^3 + A_5 q^5 + \ldots + A_v q^v.$$

Such series are known as Hermite polynomials of degree v. Note that v must be an integer as there can only be an integral number of terms in a series.

Considering equation 6.18 it may be appreciated that as q becomes larger, the exponential term becomes smaller and the Hermite polynomial becomes larger. Because the Hermite polynomial is limited to one of degree v, however, the exponential term is more powerful than the polynomial term and, on balance, ψ decays to zero as q approaches infinity. When the polynomial is limited to degree v, therefore, equation 6.18 is an acceptable solution to the Schrödinger equation. A Hermite polynomial of degree v is usually represented as $H_v(q)$ so that the solutions of the Schrödinger equation may be written as

$$\psi_v = H_v(q) \cdot e^{-\frac{1}{2}q^2} \quad (6.29)$$

When normalised, the eigenfunctions are given by

$$\psi_v = \left[\frac{\sqrt{(\alpha/\pi)}}{2^v \cdot v!} \right]^{\frac{1}{2}} H_v(q) \cdot e^{-\frac{1}{2}q^2} \quad (6.30)$$

It was pointed out above that the polynomial will be limited to degree v when

$$(2v+1-\beta) = 0$$

98 THE LINEAR HARMONIC OSCILLATOR

i.e., when
$$2v+1 = \beta \tag{6.31}$$

Substituting for β from equation 6.13
$$2v+1 = \frac{2E}{hv_e}$$

which, on rearrangement gives

$$\boxed{E_v = (v+\tfrac{1}{2})hv_e} \tag{6.32}$$

As v is the degree of the polynomial it can only have integral values and, of course, the value zero. Equation 6.32 therefore shows that the energy of the system is quantized in units of hv_e and v is called the *vibrational quantum number* which is given by

$$v = 0, 1, 2, 3, \ldots$$

It should be noted that the energy of the system can never be zero. The lowest value of the energy corresponds to the state where $v = 0$ and is given by equation 6.32 as

$$E_0 = \tfrac{1}{2}hv_e$$

which is the *zero point energy* of this system. It was pointed out in Chapter 3 that all vibratory systems have a zero point energy when the particle in a one-dimensional box was considered as a simple example of a vibratory system.

The energy levels of the linear harmonic oscillator are illustrated in *Figure 6.2* together with the probability distributions for three of the levels where the various energy levels are used as the x-axis for these functions. It should be noted that for the lowest energy level ($v = 0$) there is a maximum probability of the oscillator being at the equilibrium position. In this respect the quantum mechanical result is quite different from the classical result. With the simple harmonic motion of a mass attached to a spring, for example, the mass must be stationary at either end of its travel whereas it is moving with maximum velocity through the equilibrium position. Hence it spends the longest time at the extremities of its travel and the shortest time at the equilibrium position which is therefore the least likely place for it to be found.

Further examination of *Figure 6.2* shows that as the quantum number increases, the probability of finding the linear harmonic oscillator becomes greatest towards the extremities of its travel. This is an illustration of the correspondence principle by which the

THE LINEAR HARMONIC OSCILLATOR   99

quantum mechanical result must become identical with the classical result in the limit. It may be recalled that when representing the transition of a molecule from a lower to a higher electronic state on a potential energy diagram, the transition is drawn as a line from the mid-point of the lowest vibrational level in the lower electronic state. This is because the quantum mechanical solution above shows this to be the most probable position of the vibrator in this state.

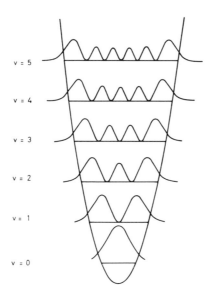

Figure 6.2   Energy levels of a linear harmonic oscillator

The simple one-dimensional box considered in Chapter 3 could be taken as a very rough approximation to the lower part of the Morse curve. *Figure 3.2* shows that the energy levels for the simple box become farther apart as the energy increases. Another approximation to the Morse curve is given by the potential function illustrated in *Figure 5.3* and once again, as shown by *Figure 5.5*, the energy levels for this system become farther apart as the energy increases. It can be seen from *Figure 6.1* that the linear harmonic oscillator is a much closer approximation to the lower part of the Morse curve and it is interesting to note that the energy levels for the oscillator are *evenly* spaced at intervals of $h\nu_e$ regardless of the total energy. This occurs because the potential energy of a harmonic

100 THE LINEAR HARMONIC OSCILLATOR

oscillator is not constant (as in the simpler models) but varies continuously and symmetrically with the x co-ordinate.

For a diatomic molecule the oscillations are *anharmonic* as can be seen from the fact that the Morse curve is not symmetrical (*Figure 6.1*). This leads to the vibrational energy levels in a real molecule becoming more closely spaced as the energy increases until they merge into a continuum when the molecule dissociates. For the energy levels of an anharmonic oscillator, equation 6.32 has to be modified by the addition of a further term to give

$$E_v = (v+\tfrac{1}{2})hv_e - (v+\tfrac{1}{2})^2 hv_e x_e \qquad (6.33)$$

where x_e is a constant known as the anharmonicity constant.

7

PARTICLE ON A RING

Up to this point all the potential energy functions which have been considered have been either constant over a linear interval or have varied only with the x co-ordinate and the only kind of momentum considered has been linear momentum. The quantization of angular momentum is extremely important in atomic and molecular systems and it is of value to consider the problem of a particle which moves around the circumference of a circle.

Consider a particle of mass, m, with constant potential energy confined to move on the circumference of a circle of radius, r. The constant value of the potential energy may be conveniently taken as zero. For this problem it is simpler to replace the cartesian co-ordinates, x, y and z, by spherical polar co-ordinates. The relationship between these two types of co-ordinates is illustrated in *Figure 7.1*. The position of the particle may be defined by values of the cartesian co-ordinates, x, y and z. Alternatively, its position may be defined by the length, r, of a line drawn to the particle from the origin together with two angles, θ and φ. The angle θ is the angle between the z-axis and the line joining the particle to the origin and φ is the angle between the x-axis and the line joining the projection of the particle in the xy plane to the origin. From *Figure 7.1* it may be seen that

$$x = r \sin \theta \cos \varphi$$
$$y = r \sin \theta \sin \varphi$$
$$z = r \cos \theta$$

The Laplacian operator, ∇^2, in terms of polar co-ordinates can be shown to be

$$\nabla^2 = \frac{\partial^2}{\partial r^2} + \frac{2}{r}\frac{\partial}{\partial r} + \frac{1}{r^2 \sin \theta}\frac{\partial}{\partial \theta}\left(\sin \theta \frac{\partial}{\partial \theta}\right) + \frac{1}{r^2 \sin^2 \theta}\frac{\partial^2}{\partial \varphi^2} \quad (7.1)$$

PARTICLE ON A RING

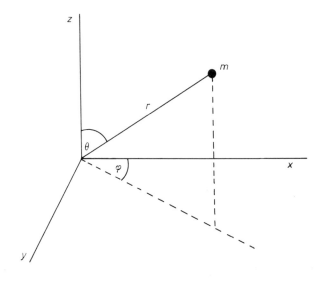

Figure 7.1 Cartesian and spherical polar co-ordinates

In the problem under examination the circle may be considered to lie in the xy plane, in which case $\theta = \pi/2$ and $\sin\theta = 1$. Moreover, as the problem involves a circle, r will be constant. As θ and r are constant for this problem, factors in the Laplacian operator such as $\partial/\partial\theta$ and $\partial/\partial r$ will be zero, so that the first three terms in equation 7.1 disappear. Putting $\sin\theta = 1$ in the remaining term reduces the Laplacian operator to

$$\nabla^2 = \frac{1}{r^2}\frac{d^2}{d\varphi^2} \tag{7.2}$$

Remembering that the three-dimensional Schrödinger equation (equation 2.71) is

$$\nabla^2\psi + \frac{8\pi^2 m}{h^2}(E - V)\psi = 0 \tag{7.3}$$

and for the current problem $V = 0$ and ∇^2 is given by equation 7.2, the appropriate form of the Schrödinger equation for the particle on a ring becomes

$$\frac{1}{r^2}\frac{d^2\psi}{d\varphi^2} + \frac{8\pi^2 m}{h^2}E\psi = 0$$

or

$$\frac{d^2\psi}{d\varphi^2} + \frac{8\pi^2 mr^2}{h^2}E\psi = 0 \tag{7.4}$$

Putting
$$\frac{8\pi^2 mr^2}{h^2} E = M^2 \quad (7.5)$$

equation 7.4 may be written in the form

$$\frac{d^2\psi}{d\varphi^2} + M^2\psi = 0 \quad (7.6)$$

Once again, because the potential energy of the system is constant, equation 7.6 is of the simpler form of the Schrödinger equation given by equation 2.74 so that the general solutions, by analogy with equations 2.75 and 2.76, are given by

and
$$\psi = A \sin M\varphi + B \cos M\varphi \quad (7.7)$$
$$\psi = C e^{iM\varphi} + D e^{-iM\varphi} \quad (7.8)$$

It was pointed out in Chapter 4 that the trigonometric expression is that of a standing wave which results from two progressive waves travelling simultaneously in opposite directions. If interest lies in the angular momentum of the particle the trigonometric expression will not be very satisfactory as it corresponds to the particle moving around the circle in both directions simultaneously. The solution in terms of complex exponentials will be much more satisfactory as it will be recalled from the discussion at the beginning of Chapter 4 that the C and D terms each correspond to motion in one direction.

The trigonometric solution does have an advantage, however, because it is a real function as opposed to a complex one and this enables the function to be plotted. Such a plot may be useful to see if any restrictions need to be placed upon the value of M in order that ψ is an eigenfunction of the system. It will be recalled that an eigenfunction must be single valued so that $\psi\psi^*$ has only one value at any point on the ring. That is to say, the eigenfunction must have the same value whenever φ is increased by 2π to $(\varphi+2\pi)$. Additionally, the eigenfunction must be continuous as also must be its first derivative, $d\psi/d\varphi$.

This problem may best be examined by considering the ring as a straight line extending from $\varphi = 0$ to $\varphi = 2\pi$ and remembering that these two limits are the same point on the ring. The sine and cosine parts of equation 7.7 may then be plotted separately and, if each part conforms to the above conditions, then equation 7.7 as a whole will be an eigenfunction of the system. These plots are illustrated in *Figure 7.2* for a few values of M.

For the value $M = \frac{1}{2}$ there is a discontinuity in the cosine function which does not have the same values at $\varphi = 0$ and $\varphi = 2\pi$. In the case of the sine function, the value of the actual function is the same at these two points, but there is a sudden change in the slope of the graph so there is a discontinuity in $d\psi/d\varphi$. From these considerations then, M cannot take the value of $\frac{1}{2}$ and a moment's reflection will show that any other fractional value is excluded by the considerations outlined above.

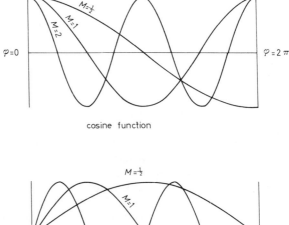

Figure 7.2 *Consideration of the allowed values of M*

For the values $M = 1$ or $M = 2$, *Figure 7.2* shows that the sine and cosine functions are single valued and continuous and that there is no discontinuity in the slopes of the graphs. Once again it will be appreciated that other integral values of M, including negative integral values, will also satisfy these conditions.

For the value $M = 0$, the sine part of equation 7.7 becomes zero but as the cosine of zero is unity the term $B \cos M\varphi$ reduces to B and the wave function has a constant value of B. This solution is therefore acceptable.

From the above discussion it may be said that equations 7.7 and

PARTICLE ON A RING   105

7.8 will be eigenfunctions of the system provided that

$$M = 0, \pm 1, \pm 2, \pm 3, \ldots$$

In view of the fact that the values of M are restricted as above, the energy of the system will be quantized. The relationship between the energy, E, of the system and the quantum number M is given by equation 7.5 from which

$$E = \frac{M^2 h^2}{8\pi^2 m r^2} \qquad (7.9)$$

Notice that as M can have the value zero, the energy of the system can be zero, hence there is no zero point rotational energy as in the case for vibrational energy. Furthermore, since E depends upon M^2, all other energy levels are doubly degenerate because E has the same value for $+M$ or $-M$.

Before leaving the trigonometric solution to the wave equation, it will be useful to normalise it as this process will provide some useful information for discussion in connection with the hydrogen atom orbitals which are discussed in the last chapter.

The condition for normalisation is

$$\int \psi \psi^* \, d\tau = 1$$

where the integral represents integration over the whole of the space in which the system exists. In the case of the particle on a ring, the variable is the angle φ which can vary from 0 to 2π. Normalisation of equation 7.7 can, therefore, be expressed in the relation

$$\int_0^{2\pi} (A \sin M\varphi + B \cos M\varphi)^2 \, d\varphi = 1$$

or

$$\int_0^{2\pi} (A^2 \sin^2 M\varphi + 2AB \sin M\varphi \cos M\varphi + B^2 \cos^2 M\varphi) \, d\varphi = 1 \quad (7.10)$$

From the trigonometric relations in Appendix 3, the following substitutions can be made.

$$\sin^2 M\varphi = \tfrac{1}{2}(1 - \cos 2M\varphi)$$
$$2 \sin M\varphi \cos M\varphi = \sin 2M\varphi$$
$$\cos^2 M\varphi = \tfrac{1}{2}(1 + \cos 2M\varphi)$$

Equation 7.10 thus becomes

$$\int_0^{2\pi} [\tfrac{1}{2}A^2(1-\cos 2M\varphi) + AB\sin 2M\varphi + \tfrac{1}{2}B^2(1+\cos 2M\varphi)]\,d\varphi = 1$$

or

$$\int_0^{2\pi} (\tfrac{1}{2}A^2 - \tfrac{1}{2}A^2\cos 2M\varphi + AB\sin 2M\varphi + \tfrac{1}{2}B^2 + \tfrac{1}{2}B^2\cos 2M\varphi)\,d\varphi = 1 \quad (7.11)$$

Integration of the sine and cosine functions between the limits 0 and 2π gives the values of zero so equation 7.11 leads to

$$A^2\pi + B^2\pi = 1 \quad (7.12)$$

Equation 7.12 is satisfied by the relationships

$$A = \frac{1}{\sqrt{\pi}}\cos\alpha \quad \text{and} \quad B = \frac{1}{\sqrt{\pi}}\sin\alpha$$

therefore equation 7.7 could be stated as

$$\psi = \frac{1}{\sqrt{\pi}}\cos\alpha\sin M\varphi + \frac{1}{\sqrt{\pi}}\sin\alpha\cos M\varphi$$

or

$$\boxed{\psi_s = \frac{1}{\sqrt{\pi}}\sin(M\varphi + \alpha)} \quad (7.13)$$

where the subscript s has been appended to ψ to indicate that the solution is in terms of a sine function.

Alternatively, equation 7.12 could equally well have been satisfied by

$$A = -\frac{1}{\sqrt{\pi}}\sin\alpha \quad \text{and} \quad B = \frac{1}{\sqrt{\pi}}\cos\alpha$$

when equation 7.7 could be stated as

$$\psi = -\frac{1}{\sqrt{\pi}}\sin\alpha\sin M\varphi + \frac{1}{\sqrt{\pi}}\cos\alpha\cos M\varphi$$

or

$$\boxed{\psi_c = \frac{1}{\sqrt{\pi}}\cos(M\varphi + \alpha)} \quad (7.14)$$

where the subscript c has been appended to ψ to indicate that the solution is in terms of a cosine function.

PARTICLE ON A RING   107

The functions ψ_s and ψ_c in equations 7.13 and 7.14 are normalised and, moreover, they are orthogonal as

$$\int_0^{2\pi} \left[\frac{1}{\sqrt{\pi}}\sin(M_1\varphi+\alpha)\right]\cdot\left[\frac{1}{\sqrt{\pi}}\cos(M_2\varphi+\alpha)\right]d\varphi = 0$$

where M_1 and M_2 are different values of the quantum number M. The functions ψ_s and ψ_c are thus orthonormal.

The term α merely shifts the origin around the ring in exactly the same manner as discussed in Chapter 2 in connection with equation 2.23. The angle α can have any value and still satisfy equation 7.12 so that there exists an infinite number of orthonormal pairs of solutions to the wave equation expressed by equations 7.13 and 7.14.

It should be pointed out in passing that the integration of equation 7.11 carried out above is not valid when $M = 0$ because integration of the trigonometric terms leads to coefficients of $1/2M$ which would be infinite if $M = 0$. To normalise the trigonometric solution to the wave equation for the special value of $M = 0$, this value has first to be inserted in equation 7.7 to give

$$\psi = B$$

which can then be normalised by writing

$$\int_0^{2\pi} B^2\, d\varphi = 1$$

This operation leads to

$$2B^2\pi = 1$$

or

$$B = \frac{1}{\sqrt{(2\pi)}}$$

so that when $M = 0$ the wave function has the constant value given by

$$\psi = \frac{1}{\sqrt{(2\pi)}}$$

The main item of interest in the problem of the particle on a ring is the angular momentum of the system. It has already been pointed out that the complex exponential solution of the wave equation will be the more useful in this context as the C and D terms in equation 7.8 each correspond to motion in one direction only. From this point of view it is best to restrict attention, for the time being, to positive values of M, separate the two terms and write

$$\psi_+ = C\,e^{iM\varphi} \qquad (7.15)$$

108 PARTICLE ON A RING

and
$$\psi_- = D e^{-iM\varphi} \tag{7.16}$$

where the *subscripts* refer to the sign of the power to which the exponential is raised and should not be confused with the *superscripts* which indicated symmetry and antisymmetry in Chapter 5.

Normalisation of the wave function given by equation 7.15 may be achieved by putting

$$\int_0^{2\pi} \psi_+ \psi_+^* \, d\varphi = 1$$

$$\int_0^{2\pi} (C e^{iM\varphi} \cdot C^* e^{-iM\varphi}) \, d\varphi = 1$$

$$CC^* \int_0^{2\pi} d\varphi = 1$$

$$CC^* = \frac{1}{2\pi}$$

whence
$$C = \frac{1}{\sqrt{(2\pi)}}$$

Similar normalisation of equation 7.16 shows that

$$D = \frac{1}{\sqrt{(2\pi)}}$$

so that the normalised forms of equations 7.15 and 7.16 are

$$\boxed{\psi_+ = \frac{1}{\sqrt{(2\pi)}} e^{iM\varphi}} \tag{7.17}$$

$$\boxed{\psi_- = \frac{1}{\sqrt{(2\pi)}} e^{-iM\varphi}} \tag{7.18}$$

These solutions are now normalised and they are also orthogonal as

$$\int_0^{2\pi} \left(\frac{1}{\sqrt{(2\pi)}} e^{iM_1\varphi}\right) \left(\frac{1}{\sqrt{(2\pi)}} e^{-iM_2\varphi}\right) d\varphi = 0$$

If desired, the origin on the ring may be shifted by an angle α when equation 7.17, for example, would be written

$$\psi_+ = \frac{1}{\sqrt{(2\pi)}} e^{i(M\varphi + \alpha)}$$

PARTICLE ON A RING 109

For the sake of simplicity in the rest of the discussion, α will be set equal to zero.

It will now be appreciated that if M has the value $+1$, say, in equation 7.7, this equation becomes identical with equation 7.18 in which M has a value of -1. From the point of view of angular momentum determination then, equations 7.17 and 7.18 could be combined as

$$\psi_\pm = \frac{1}{\sqrt{(2\pi)}} e^{iM\varphi} \tag{7.19}$$

in which the possibility of M having positive or negative values is taken into account.

The ring on which the particle moves has been considered to lie in the xy plane so that the particle will have an angular momentum about the z-axis. Using the wave functions expressed by equation 7.19 the angular momentum should be a sharp quantity as any value of M corresponds only to motion in one direction. It should thus be possible to determine the angular momentum of the particle by operating on $\psi_\pm$ with the angular momentum operator.

The operator for angular momentum about the z-axis is given in polar co-ordinates by $(h/2\pi i)(d/d\varphi)$. Operating on the function $\psi_\pm$ with this operator gives

$$\frac{h}{2\pi i} \cdot \frac{d}{d\varphi}(\psi_\pm) = \frac{h}{2\pi i} \cdot \frac{d}{d\varphi}\left(\frac{1}{\sqrt{(2\pi)}} e^{iM\varphi}\right)$$

$$= \frac{h}{2\pi i} \cdot iM \cdot \frac{1}{\sqrt{(2\pi)}} e^{iM\varphi}$$

$$= \frac{Mh}{2\pi} \psi_\pm$$

This operation has yielded a real number multiplied by the original wave function so that the real number is the angular momentum of the particle. Denoting the angular momentum about the z-axis as L_z, then

$$L_z = \frac{Mh}{2\pi}$$

Notice that the angular momentum about the z-axis is quantized in units of $h/2\pi$ which was an arbitrary assumption in the Bohr theory of the hydrogen atom.

The angular momentum of the particle can be zero as M may take the value zero. Positive values of M in the above equation will lead to positive values of the angular momentum which corresponds to motion around the ring in one particular direction. Conversely, negative values of M lead to negative values of angular momentum which correspond to motion around the ring in the opposite direction.

One further point may be illustrated by the problem of the particle on the ring. The wave functions represented by equation 7.19 cannot be plotted to form a diagram as they contain an imaginary part. The trigonometric functions as given by equations 7.13 and 7.14 do not suffer from this disadvantage. In fact, the trigonometric wave functions are linear combinations of the exponential functions. Considering equation 7.13 with $\alpha = 0$

$$\psi_s = \frac{1}{\sqrt{\pi}} \sin M\varphi \tag{7.20}$$

$$= \frac{1}{2i} \cdot \frac{1}{\sqrt{\pi}} (e^{iM\varphi} - e^{-iM\varphi})$$

$$= \frac{1}{i\sqrt{2}} \left(\frac{1}{\sqrt{(2\pi)}} e^{iM\varphi} - \frac{1}{\sqrt{(2\pi)}} e^{-iM\varphi} \right)$$

or

$$\boxed{\psi_s = \frac{1}{i\sqrt{2}} (\psi_+ - \psi_-)} \tag{7.21}$$

Again, considering equation 7.14 with $\alpha = 0$

$$\psi_c = \frac{1}{\sqrt{\pi}} \cos M\varphi \tag{7.22}$$

$$= \frac{1}{2} \cdot \frac{1}{\sqrt{\pi}} (e^{iM\varphi} + e^{-iM\varphi})$$

$$= \frac{1}{\sqrt{2}} \left(\frac{1}{\sqrt{(2\pi)}} e^{iM\varphi} + \frac{1}{\sqrt{(2\pi)}} e^{-iM\varphi} \right)$$

or

$$\boxed{\psi_c = \frac{1}{\sqrt{2}} (\psi_+ + \psi_-)} \tag{7.23}$$

These linear combinations illustrate a general property of degenerate wave functions which is that any linear combination of the members of a degenerate set is also an acceptable wave function and has the same energy as the original functions. In equations 7.21 and 7.23 above, ψ_+ and ψ_- for a given value of M will be degenerate and linear combinations of them give ψ_s and ψ_c which are already known to be acceptable wave functions and which have the same energy as ψ_+ and ψ_- for the same value of M.

8

THE RIGID ROTATOR

Having initiated the study of a rotational problem in the last chapter, it is instructive to consider a problem which serves as a model for the rotational spectra of diatomic molecules. Such a model is the rigid rotator in which two masses, separated by a fixed distance, rotate about their common centre of mass. Additionally, some aspects of this problem are relevant to that of the hydrogen atom which is considered in the next chapter.

Consider two masses, m_1 and m_2, distant r_1 and r_2 respectively from their common centre of mass, rotating about the centre of mass. It is convenient to take the centre of mass as the origin of a co-ordinate system when the rigid rotator may be represented as shown in *Figure 8.1*.

The velocity, v, of a single particle may be written in terms of its component velocities, v_x, v_y and v_z parallel to the cartesian axes as

$$v^2 = v_x^2 + v_y^2 + v_z^2 \tag{8.1}$$

or, remembering that

$$v_x = \frac{dx}{dt}, \quad v_y = \frac{dy}{dt} \quad \text{and} \quad v_z = \frac{dz}{dt}$$

and writing the differentials of x, y and z with respect to time as $\dot{x}$, $\dot{y}$ and $\dot{z}$, equation 8.1 may be written

$$v^2 = \dot{x}^2 + \dot{y}^2 + \dot{z}^2 \tag{8.2}$$

The kinetic energy, T, of a single particle is thus given by

$$T = \tfrac{1}{2}mv^2 = \tfrac{1}{2}m(\dot{x}^2 + \dot{y}^2 + \dot{z}^2) \tag{8.3}$$

For rotational problems it is simpler to use the spherical polar co-ordinates, r, θ, and φ. Equation 8.3 may be transformed into polar co-ordinates by a straightforward but rather lengthy argument when it provides the result

$$T = \tfrac{1}{2}m(\dot{r}^2 + r^2\dot{\theta}^2 + r^2\dot{\varphi}^2 \sin^2 \theta) \qquad (8.4)$$

where the dotted symbols once again represent a differential with respect to time.

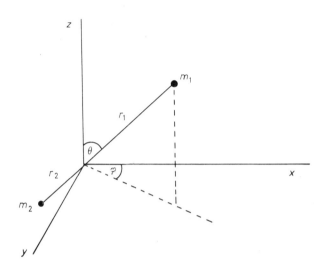

Figure 8.1 The rigid rotator

If the distance of the particle from the origin is fixed, r will be constant and hence $\dot{r} = 0$. Under these circumstances equation 8.4 becomes

$$T = \tfrac{1}{2}m(r^2\dot{\theta}^2 + r^2\dot{\varphi}^2 \sin^2 \theta)$$

or

$$T = \tfrac{1}{2}mr^2(\dot{\theta}^2 + \dot{\varphi}^2 \sin^2 \theta) \qquad (8.5)$$

For the two particles represented in *Figure 8.1*, the total kinetic energy of the system will be the sum of the individual kinetic energies of the particles so that from equation 8.5

$$T = \tfrac{1}{2}m_1 r_1^2(\dot{\theta}^2 + \dot{\varphi}^2 \sin^2 \theta) + \tfrac{1}{2}m_2 r_2^2(\dot{\theta}^2 + \dot{\varphi}^2 \sin^2 \theta) \qquad (8.6)$$

In equation 8.6 the subscripts 1 and 2 have been used to represent the two particles in respect of their masses and distances from the origin. The angular dependence of motion of the two particles, however, must be identical so that the values of the θ and φ terms for

THE RIGID ROTATOR 113

each particle are the same. Equation 8.6 may thus be written

$$T = \tfrac{1}{2}(m_1 r_1{}^2 + m_2 r_2{}^2)(\dot{\theta}^2 + \dot{\varphi}^2 \sin^2 \theta) \tag{8.7}$$

Now, the moment of inertia, I, of a system is defined as

$$I = \Sigma m_i r_i{}^2$$

so that in the case of the rigid rotator

$$I = (m_1 r_1{}^2 + m_2 r_2{}^2)$$

and equation 8.7 may be stated as

$$T = \tfrac{1}{2} I(\dot{\theta}^2 + \dot{\varphi}^2 \sin^2 \theta) \tag{8.8}$$

Comparison of equation 8.8 for the rigid rotator with equation 8.5 for a single particle shows that the rigid rotator behaves as though it were a single particle of mass I at unit distance from the origin.

It will be recalled that the three-dimensional Schrödinger equation for a single particle is

$$\nabla^2 \psi + \frac{8\pi^2 m}{h^2}(E - V)\psi = 0 \tag{8.9}$$

The potential energy of the rigid rotator will be constant and this constant value may be conveniently taken as zero. Applying equation 8.9 to the rigid rotator and putting $V = 0$ and $m = I$ gives

$$\nabla^2 \psi + \frac{8\pi^2 I}{h^2} E\psi = 0 \tag{8.10}$$

The Laplacian operator was given in terms of spherical polar coordinates by equation 7.1 as

$$\nabla^2 = \frac{\partial^2}{\partial r^2} + \frac{2}{r}\frac{\partial}{\partial r} + \frac{1}{r^2 \sin \theta}\frac{\partial}{\partial \theta}\left(\sin \theta \frac{\partial}{\partial \theta}\right) + \frac{1}{r^2 \sin^2 \theta}\frac{\partial^2}{\partial \varphi^2} \tag{8.11}$$

For the rigid rotator, r has the constant value of unity. Since r is constant, factors involving $\partial/\partial r$ will be zero and putting $r = 1$ in the remaining terms reduces the Laplacian operator so that equation 8.10 takes the form

$$\frac{1}{\sin \theta}\frac{\partial}{\partial \theta}\left(\sin \theta \frac{\partial \psi}{\partial \theta}\right) + \frac{1}{\sin^2 \theta}\frac{\partial^2 \psi}{\partial \varphi^2} + \frac{8\pi^2 I}{h^2} E\psi = 0 \tag{8.12}$$

This equation contains two variables, θ and φ, and the approach to its solution is analogous to the approach used for the three-dimensional box in Chapter 3, where an attempt is made to separate the variables.

It is assumed that ψ is a product of two functions, each of which is a function of only one of the variables. It is assumed, therefore,

that
$$\psi(\theta, \varphi) = T(\theta) \cdot F(\varphi) \tag{8.13}$$

which means that ψ (which is a function of θ and φ) is equal to the product of two functions T and F where T is a function only of θ and F is a function only of φ. Equation 8.13 may be written more simply as

$$\psi = T \cdot F \tag{8.14}$$

Since the function F is independent of θ, differentiation of equation 8.14 with respect to θ yields

$$\frac{\partial \psi}{\partial \theta} = F \cdot \frac{dT}{d\theta} \tag{8.15}$$

Similarly, differentiation with respect to φ gives

$$\frac{\partial \psi}{\partial \varphi} = T \cdot \frac{dF}{d\varphi}$$

and further differentiation with respect to φ yields

$$\frac{\partial^2 \psi}{\partial \varphi^2} = T \cdot \frac{d^2 F}{d\varphi^2} \tag{8.16}$$

Substituting from equations 8.14, 8.15 and 8.16 into equation 8.12 gives

$$\frac{F}{\sin \theta} \frac{d}{d\theta}\left(\sin \theta \frac{dT}{d\theta}\right) + \frac{T}{\sin^2 \theta} \frac{d^2 F}{d\varphi^2} + \frac{8\pi^2 I}{h^2} ETF = 0$$

Multiplying by $\sin^2 \theta / TF$

$$\frac{\sin \theta}{T} \frac{d}{d\theta}\left(\sin \theta \frac{dT}{d\theta}\right) + \frac{1}{F} \frac{d^2 F}{d\varphi^2} + \frac{8\pi^2 I}{h^2} E \sin^2 \theta = 0$$

or

$$\frac{\sin \theta}{T} \frac{d}{d\theta}\left(\sin \theta \frac{dT}{d\theta}\right) + \frac{8\pi^2 I}{h^2} E \sin^2 \theta = -\frac{1}{F} \frac{d^2 F}{d\varphi^2} \tag{8.17}$$

Each side of equation 8.17 contains only one variable. The left-hand side has only terms in the variable θ and the right-hand side has only a term in the variable φ. As the equation must hold for all values of θ and φ, each side of the equation must be a constant. Representing this constant value by m^2 allows each side of equation 8.17 to be written

$$\boxed{-\frac{1}{F} \frac{d^2 F}{d\varphi^2} = m^2} \tag{8.18}$$

and

$$\boxed{\frac{\sin\theta}{T}\frac{d}{d\theta}\left(\sin\theta\frac{dT}{d\theta}\right)+\frac{8\pi^2 I}{h^2}E\sin^2\theta = m^2}$$ (8.19)

That this separation is possible justifies the original assumption expressed in equation 8.13.

If equations 8.18 and 8.19 can be solved to give acceptable functions T and F, then the wave function, ψ, is given by equation 8.13 as the product of these two functions. Equations 8.18 and 8.19 may be called the F and T equations respectively and their solutions are considered separately below.

THE F EQUATION

The F equation may be written in the form

$$\frac{1}{F}\frac{d^2 F}{d\varphi^2}+m^2 = 0$$

or

$$\frac{d^2 F}{d\varphi^2}+m^2 F = 0$$ (8.20)

This equation is of exactly the same form as that for a particle on a ring given by equation 7.6 except that m^2 appears instead of M^2 and F appears instead of ψ.

The eigenfunctions of the F equation are thus of exactly the same form as those of equation 7.6 in the last chapter and they may be expressed as

$$F_s = \frac{1}{\sqrt{\pi}}\sin m\varphi$$ (8.21)

and

$$F_c = \frac{1}{\sqrt{\pi}}\cos m\varphi$$ (8.22)

or, in terms of complex exponentials,

$$F_+ = \frac{1}{\sqrt{(2\pi)}}e^{im\varphi}$$ (8.23)

and

$$F_- = \frac{1}{\sqrt{(2\pi)}}e^{-im\varphi}$$ (8.24)

where
$$m = 0, \pm 1, \pm 2, \pm 3, \ldots \qquad (8.25)$$

THE T EQUATION

The solution of the F equation above has generated a quantum number, m, and the solution of the T equation generates a further quantum number which has a relationship with m. In order to clarify this relationship it may be helpful to anticipate the result of the solution of the T equation. The solution to the T equation is a function known as an *associated Legendre polynomial*, but before considering this function it is advisable to define a Legendre polynomial.

A Legendre polynomial of degree l in the variable q is usually represented as $P_l(q)$ and is obtained by differentiating the expression $(q^2 - 1)^l$ l times and then dividing the result by $2^l \cdot l!$ Thus

$$P_l(q) = \frac{1}{2^l \cdot l!} \frac{d^l (q^2 - 1)^l}{dq^l} \qquad (8.26)$$

An associated Legendre polynomial is obtained by differentiating the corresponding Legendre polynomial a further m times and then multiplying the result by $(1 - q^2)^{m/2}$. Representing the associated Legendre polynomial by $P_l^m(q)$ then,

$$P_l^m(q) = (1 - q^2)^{m/2} \cdot \frac{d^m P_l(q)}{dq^m} \qquad (8.27)$$

It should be noted that both l and m must be integers as they each represent a number of differentiations. Moreover, m cannot be greater than l. This follows from equation 8.26 which shows that the highest power of q in the polynomial $P_l(q)$ is q^l. For example, suppose $l = 2$, then equation 8.26 becomes

$$\begin{aligned}
P_2(q) &= \frac{1}{2^2 \cdot 2!} \frac{d^2 (q^2 - 1)^2}{dq^2} \\
&= \frac{1}{8} \frac{d^2}{dq^2} (q^4 - 2q^2 + 1) \\
&= \frac{1}{8} \frac{d}{dq} (4q^3 - 4q) \\
&= \tfrac{1}{8}(12q^2 - 4) \\
&= \tfrac{3}{2} q^2 - \tfrac{1}{2}
\end{aligned}$$

Thus, if the highest power of q is q^l, differentiation of $P_l(q)$ m times

THE RIGID ROTATOR 117

reduces the highest power of q to q^{l-m} so that if $m > l$, the associated Legendre polynomial, $P_l^m(q)$, vanishes.

With the above preliminary remarks in mind, the solution of the T equation may be considered.

Putting

$$\frac{8\pi^2 I}{h^2} E = \beta \tag{8.28}$$

in equation 8.19, the T equation becomes

$$\frac{\sin\theta}{T} \frac{d}{d\theta}\left(\sin\theta \frac{dT}{d\theta}\right) + \beta \sin^2\theta - m^2 = 0$$

Multiplying by $T/\sin^2\theta$

$$\frac{1}{\sin\theta} \frac{d}{d\theta}\left(\sin\theta \frac{dT}{d\theta}\right) + \left(\beta - \frac{m^2}{\sin^2\theta}\right)T = 0$$

Carrying out the differentiation indicated in the first term and remembering that $(\sin\theta).(dT/d\theta)$ must be differentiated as a product,

$$\frac{1}{\sin\theta}\left(\sin\theta \frac{d^2 T}{d\theta^2} + \cos\theta \frac{dT}{d\theta}\right) + \left(\beta - \frac{m^2}{\sin^2\theta}\right)T = 0 \tag{8.29}$$

The variable in the equation is now changed by putting

$$q = \cos\theta \tag{8.30}$$

Now

$$\frac{dT}{d\theta} = \frac{dT}{dq} \cdot \frac{dq}{d\theta} = \frac{dT}{dq}(-\sin\theta) \tag{8.31}$$

Furthermore,

$$\frac{d^2 T}{d\theta^2} = \frac{d}{d\theta}\left(\frac{dT}{d\theta}\right)$$

Substituting from equation 8.31

$$\frac{d^2 T}{d\theta^2} = \frac{d}{d\theta}\left(-\sin\theta \cdot \frac{dT}{dq}\right)$$

and differentiating $(\sin\theta).(dT/dq)$ as a product

$$\frac{d^2 T}{d\theta^2} = -\sin\theta \cdot \frac{d}{d\theta}\left(\frac{dT}{dq}\right) + \frac{dT}{dq}(-\cos\theta) \tag{8.32}$$

Remembering that

$$\frac{d}{d\theta} = \frac{d}{dq} \cdot \frac{dq}{d\theta}$$

for the first term of equation 8.32, the equation may be written

$$\frac{d^2T}{d\theta^2} = -\sin\theta\left(\frac{d^2T}{dq^2}\cdot\frac{dq}{d\theta}\right) + \frac{dT}{dq}(-\cos\theta) \tag{8.33}$$

Since q is defined by equation 8.30, $dq/d\theta$ is given by

$$\frac{dq}{d\theta} = -\sin\theta$$

and substituting in equation 8.33

$$\frac{d^2T}{d\theta^2} = \sin^2\theta\cdot\frac{d^2T}{dq^2} - \cos\theta\cdot\frac{dT}{dq} \tag{8.34}$$

The expressions for $dT/d\theta$ and $d^2T/d\theta^2$ given by equations 8.31 and 8.34 may now be substituted in equation 8.29 to give

$$\frac{1}{\sin\theta}\left[\sin\theta\left(\sin^2\theta\cdot\frac{d^2T}{dq^2} - \cos\theta\cdot\frac{dT}{dq}\right) - \sin\theta\cos\theta\cdot\frac{dT}{dq}\right] +$$
$$+ \left(\beta - \frac{m^2}{\sin^2\theta}\right)T = 0$$

or

$$\sin^2\theta\cdot\frac{d^2T}{dq^2} - 2\cos\theta\cdot\frac{dT}{dq} + \left(\beta - \frac{m^2}{\sin^2\theta}\right)T = 0 \tag{8.35}$$

Finally, remembering that $q = \cos\theta$ and hence $\sin^2\theta = 1 - q^2$,

$$(1-q^2)\frac{d^2T}{dq^2} - 2q\cdot\frac{dT}{dq} + \left(\beta - \frac{m^2}{1-q^2}\right)T = 0 \tag{8.36}$$

If, now, the quantity β is represented by putting

$$\beta = l(l+1) \tag{8.37}$$

equation 8.36 takes the form

$$(1-q^2)\frac{d^2T}{dq^2} - 2q\cdot\frac{dT}{dq} + \left[l(l+1) - \frac{m^2}{1-q^2}\right]T = 0 \tag{8.38}$$

which is an equation known as an associated Legendre equation. The solutions for this equation are the associated Legendre polynomials $P_l^m(q)$ of degree l and order m which were defined by equation 8.27.

It will be recalled from the discussion of associated Legendre polynomials that m cannot be greater than l. Furthermore, the number m corresponds to a number of differentiation operations. The solution of the F equation showed that m could have negative as well as positive values, but a function cannot be differentiated a negative number of times so that the functions $T(\theta)$ which are

THE RIGID ROTATOR   119

solutions to the T equation may be represented

$$T(\theta) = P_l^{|m|}(q) \tag{8.39}$$

where $|m|$, the modulus of m (i.e. its magnitude regardless of sign), is written instead of m. The quantum number l arises only from the associated Legendre polynomial and has therefore only positive values.

When considering the rotational spectra of diatomic molecules, the quantum number is usually written as J rather than l. Additionally, $q = \cos\theta$ so that for rotational spectra purposes, equation 8.39 is usually written in the form

$$T(\theta) = P_J^{|m|}(\cos\theta) \tag{8.40}$$

where

$$J = 0, 1, 2, 3, \ldots$$

and

$$m = -J, -(J-1), \ldots, -1, 0, +1, \ldots, +J$$

Equation 8.40 may be normalised to yield

$$T(\theta) = \sqrt{\left[\frac{(2J+1)(J-|m|)!}{2(J+|m|)!}\right]} P_J^{|m|}(\cos\theta) \tag{8.41}$$

THE ENERGY LEVELS

The constant β is related to the quantum number J by equation 8.37, where l was originally written in place of J,

$$\beta = J(J+1) \tag{8.42}$$

In the first instance, however, β was related to the energy of the system by equation 8.28

$$\beta = \frac{8\pi^2 I}{h^2} E \tag{8.28}$$

Hence, from equations 8.28 and 8.42

$$\boxed{E = \frac{h^2}{8\pi^2 I} \cdot J(J+1)} \tag{8.43}$$

This relationship gives the eigenvalues of the energy of the rotator and J is known as the rotational quantum number.

The state of the system, however, requires the specification of the two quantum numbers J and m. It will be remembered that the relationship between these two quantum numbers is $J \geqslant |m|$. When $J = 2$ for example, then $m = +2, +1, 0, -1, -2$. There are thus five possible states of the system for $J = 2$. As the energy is determined solely by the value of J in equation 8.43 there are thus five different states of the system with the same energy. For $J = 2$ then, there exists fivefold degeneracy of the system. In general, for any value of J there are $(2J+1)$ degenerate states. It may be mentioned here that this degeneracy is removed if the molecule is placed in a magnetic field and hence, in the presence of a magnetic field extra lines appear in the rotational spectrum. It should be noted that J can have the value zero, therefore it is once again seen that a rotational system can have zero energy.

The rigid rotator serves as an approximate model for the rotational spectra of diatomic molecules. If two energy levels are defined by the rotational quantum numbers J and J' then the energy difference, ΔE, between them is given by equation 8.43 as

$$\Delta E = \frac{h^2}{8\pi^2 I}[J'(J'+1) - J(J+1)] \qquad (8.44)$$

There is a selection rule in rotational spectroscopy that only permits transitions between two states for the case where $\Delta J = \pm 1$. That is to say, transitions can only occur between successive energy levels. To obtain a value of ΔE for this case it is only necessary to put $J' = J+1$ in equation 8.44 when

$$\Delta E = \frac{h^2}{8\pi^2 I}[(J+1)(J+2) - J(J+1)]$$

or

$$\Delta E = \frac{h^2}{8\pi^2 I} \cdot 2(J+1) \qquad (8.45)$$

The frequency, v, of the spectral line corresponding to this transition is given by equation 1.2 as

$$v = \Delta E/h$$

or, substituting for ΔE from equation 8.45,

$$v = \frac{h}{8\pi^2 I} \cdot 2(J+1) \qquad (8.46)$$

The frequencies of the spectral lines for values of $J = 0, 1, 2, 3$, etc., are given in units of $h/8\pi^2 I$ by equation 8.46 as 2, 4, 6, 8, etc. There is a constant difference of $2h/8\pi^2 I$ between successive frequencies so that the lines in the rotational spectrum should be evenly spaced.

In fact, a diatomic molecule is not a *rigid* rotator. The bond between the atoms can be compressed or extended so that a diatomic molecule should strictly be regarded as a *non-rigid* rotator. The rotation of the molecule tends to stretch the bond between the atoms and the greater the energy of rotation (i.e. the greater the value of J), the greater is the degree of extension of the bond. This has an effect on the energy levels, as given by equation 8.43, which has to be modified to

$$E = \frac{h^2}{8\pi^2 I} \cdot J(J+1) - \frac{h^4}{32\pi^4 I^2 k r^2} \cdot J^2(J+1)^2 \qquad (8.47)$$

where r is the length of the bond in the absence of rotation and k is the force constant for the centrifugal stretching of the bond.

The energy levels of the non-rigid rotator are more closely spaced than those of a rigid rotator and as a result the spacing between the rotational spectral lines of a real molecule decreases as J increases.

9

THE HYDROGEN ATOM

It is convenient to treat the hydrogen atom and the hydrogen-like ions, He^+, Li^{2+}, etc., together since they differ only in nuclear charge. The potential energy of the electron in a hydrogen-like species has already been given in Chapter 5 by equation 5.27 which may be stated here in the form

$$V = -\frac{Ze^2}{4\pi\varepsilon_0 r} \qquad (9.1)$$

where r is the distance of the electron from the nucleus in any direction, Z is the atomic number and e is the electronic charge.

The Schrödinger equation for this system thus takes the form

$$\nabla^2 \psi + \frac{8\pi^2 m}{h^2}\left(E + \frac{Ze^2}{4\pi\varepsilon_0 r}\right)\psi = 0 \qquad (9.2)$$

Strictly, the reduced mass of the nucleus and the electron should be used instead of m, the mass of the electron. The nucleus is so massive, however, that there is no significant difference between these two quantities.

Expressing the Laplacian operator in spherical polar co-ordinates, equation 9.2 takes the form

$$\frac{1}{r^2}\frac{\partial}{\partial r}\left(r^2 \frac{\partial \psi}{\partial r}\right) + \frac{1}{r^2 \sin \theta}\frac{\partial}{\partial \theta}\left(\sin \theta \frac{\partial \psi}{\partial \theta}\right) + \frac{1}{r^2 \sin^2 \theta}\frac{\partial^2 \psi}{\partial \varphi^2}$$
$$+ \frac{8\pi^2 m}{h^2}\left(E + \frac{Ze^2}{4\pi\varepsilon_0 r}\right)\psi = 0 \qquad (9.3)$$

where the first two terms of the expression for the Laplacian

THE HYDROGEN ATOM

operator, first given by equation 7.1, have been combined since

$$\frac{1}{r^2}\frac{\partial}{\partial r}\left(r^2\frac{\partial \psi}{\partial r}\right) = \frac{1}{r^2}\left(r^2\frac{\partial^2 \psi}{\partial r^2} + 2r\frac{\partial \psi}{\partial r}\right)$$

$$= \frac{\partial^2 \psi}{\partial r^2} + \frac{2}{r}\frac{\partial \psi}{\partial r}$$

The procedure for solving equation 9.3 is analogous to that used in the rigid rotator problem, but in this case r is a variable so that ψ must be regarded as a function of the three variables, r, θ and φ. It is assumed initially that the variables may be separated and that ψ may be regarded as the product of three functions, each of which is a function of only one of the variables. It is assumed, therefore, that

$$\psi(r, \theta, \varphi) = R(r) \cdot T(\theta) \cdot F(\varphi) \tag{9.4}$$

where R is a function only of r, T is a function only of θ, and F is a function only of φ. Equation 9.4 may be written more simply as

$$\psi = RTF \tag{9.5}$$

Substituting for ψ in equation 9.3 and then dividing throughout by RTF gives

$$\frac{1}{r^2 R}\frac{d}{dr}\left(r^2\frac{dR}{dr}\right) + \frac{1}{r^2 \sin\theta}\frac{1}{T}\frac{d}{d\theta}\left(\sin\theta\frac{dT}{d\theta}\right) + \frac{1}{r^2 \sin^2\theta}\frac{1}{F}\frac{d^2 F}{d\varphi^2}$$

$$+ \frac{8\pi^2 m}{h^2}\left(E + \frac{Ze^2}{4\pi\varepsilon_0 r}\right) = 0$$

Multiplying by $r^2 \sin^2\theta$ and rearranging

$$\frac{\sin^2\theta}{R}\frac{d}{dr}\left(r^2\frac{dR}{dr}\right) + \frac{\sin\theta}{T}\frac{d}{d\theta}\left(\sin\theta\frac{dT}{d\theta}\right)$$

$$+ \frac{8\pi^2 m}{h^2}\left(E + \frac{Ze^2}{4\pi\varepsilon_0 r}\right)r^2 \sin^2\theta = -\frac{1}{F}\frac{d^2 F}{d\varphi^2} \tag{9.6}$$

The right-hand side of equation 9.6 contains only one variable and since the equation must hold for all values of r, θ, and φ, each side of the equation must be constant. Representing this constant by m^2 allows each side of equation 9.6 to be written

$$\frac{1}{F}\frac{d^2 F}{d\varphi^2} = -m^2 \tag{9.7}$$

124 THE HYDROGEN ATOM

and

$$\frac{\sin^2\theta}{R}\frac{d}{dr}\left(r^2\frac{dR}{dr}\right) + \frac{\sin\theta}{T}\frac{d}{d\theta}\left(\sin\theta\frac{dT}{d\theta}\right) +$$
$$+ \frac{8\pi^2 m}{h^2}\left(E + \frac{Ze^2}{4\pi\varepsilon_0 r}\right)r^2\sin^2\theta = m^2 \quad (9.8)$$

Equation 9.8 still contains two of the variables but these may be separated by dividing the equation by $\sin^2\theta$ and rearranging to give

$$\frac{1}{R}\frac{d}{dr}\left(r^2\frac{dR}{dr}\right) + \frac{8\pi^2 m}{h^2}\left(E + \frac{Ze^2}{4\pi\varepsilon_0 r}\right)r^2$$
$$= \frac{m^2}{\sin^2\theta} - \frac{1}{T\sin\theta}\frac{d}{d\theta}\left(\sin\theta\frac{dT}{d\theta}\right) \quad (9.9)$$

Each side of equation 9.9 contains only one variable and, because the equation must hold for all values of r and θ, each side of the equation must have a constant value. Representing this constant by β, it follows that

$$\frac{m^2}{\sin^2\theta} - \frac{1}{T\sin\theta}\frac{d}{d\theta}\left(\sin\theta\frac{dT}{d\theta}\right) = \beta \quad (9.10)$$

and

$$\frac{1}{R}\frac{d}{dr}\left(r^2\frac{dR}{dr}\right) + \frac{8\pi^2 m}{h^2}\left(E + \frac{Ze^2}{4\pi\varepsilon_0 r}\right)r^2 = \beta \quad (9.11)$$

It has thus been possible to separate the three variables in equation 9.3 and to write three separate equations 9.7, 9.10 and 9.11 for each variable. The original assumption expressed in equation 9.4 is thus justified and the problem is reduced to solving the three separate equations. These equations may be written in the following form:

$$\boxed{\frac{1}{F}\frac{d^2 F}{d\varphi^2} = -m^2} \quad (9.12)$$

$$\boxed{\frac{1}{\sin\theta}\frac{d}{d\theta}\left(\sin\theta\frac{dT}{d\theta}\right) + \left(\beta - \frac{m^2}{\sin^2\theta}\right)T = 0} \quad (9.13)$$

$$\boxed{\frac{1}{r^2}\frac{d}{dr}\left(r^2\frac{dR}{dr}\right) + \left[\frac{8\pi^2 m}{h^2}\left(E + \frac{Ze^2}{4\pi\varepsilon_0 r}\right) - \frac{\beta}{r^2}\right]R = 0} \quad (9.14)$$

THE HYDROGEN ATOM

Equations 9.12, 9.13 and 9.14 may be called respectively, the F, T and R equations.

THE F EQUATION

The F equation is exactly the same as that encountered in the rigid rotator problem (equation 8.20) and both are, of course, the same as the equation for the particle on a ring (equation 7.6). The solutions of the F equation in the present context are therefore the same as for the F equation in the rigid rotator and the equation of the particle on a ring. These solutions may be expressed in the form of trigonometric functions or in the form of complex exponentials. It may be recalled from the discussion of the particle on a ring that if interest lies in the angular momentum of the particle, the complex exponential solution is the more appropriate. It is shown later that the quantum number derived from the F equation is associated with a component of the angular momentum of the electron in a hydrogen atom, so it is convenient, for the time being, to express the solution of the F equation in terms of complex exponentials. The solutions may thus be stated

$$F = \frac{1}{\sqrt{(2\pi)}} e^{im\varphi} \qquad (9.15)$$

where

$$m = 0, \pm 1, \pm 2, \pm 3, \ldots \qquad (9.16)$$

THE T EQUATION

Once again the T equation is of the same form as the T equation for the rigid rotator (equation 8.29) and will therefore have solutions of the same form. These solutions involve the quantum number m and a further quantum number which is denoted by l in the case of electronic motion as opposed to J for rotational spectra. The quantum number l is thus related to the quantity β in equation 9.13 by

$$\beta = l(l+1) \qquad (9.17)$$

The permitted values of l are given by

$$l = 0, 1, 2, 3, \ldots \qquad (9.18)$$

and there is the further restriction that

$$|m| \leq l \qquad (9.19)$$

THE R EQUATION

The solution of the R equation generates a further quantum number, n, which has a relationship with the quantum number l. It will be appreciated that as the R equation (equation 9.14) contains the quantity β which is given by equation 9.17 the quantum number l will be involved in the solution of the R equation. The solution of the R equation is a function known as an *associated Laguerre polynomial* and a consideration of this function will illustrate the relationship between the quantum numbers n and l.

A Laguerre polynomial of degree k in the variable q may be represented as $L_k(q)$ and is obtained by differentiating the expression $q^k e^{-q}$, k times and then multiplying the result by e^q. Thus

$$L_k(q) = e^q \cdot \frac{d^k(q^k e^{-q})}{dq^k} \qquad (9.20)$$

An associated Laguerre polynomial is obtained by differentiating the corresponding Laguerre polynomial s times. Representing the associated Laguerre polynomial by $L_k^s(q)$ then,

$$L_k^s(q) = \frac{d^s L_k(q)}{dq^s} \qquad (9.21)$$

In the above discussion it will be appreciated that both k and s are required to be positive integers since they each represent a number of differentiating operations. The number k may take the value of zero when $L_0(q) = 1$. From equation 9.20 it may be seen that the highest power of q in the polynomial $L_k(q)$ is q^k. Differentiation of $L_k(q)$, s times reduces the highest power of q to q^{k-s} so that if $s > k$ the associated Laguerre polynomial $L_k^s(q)$ vanishes. The value of s must be less than or equal to k so that the restrictions on the values of k and s may be stated

$$k = 0, 1, 2, 3, \ldots \qquad (9.22)$$

$$s \leqslant k \qquad (9.23)$$

In the solution of the R equation the associated Laguerre polynomials take the form

$$L_{n+1}^{2l+1}\left(\frac{2Zr}{na_0}\right)$$

where a_0 is given by

$$a_0 = \frac{(4\pi\varepsilon_0)h^2}{4\pi^2 m e^2}$$

and is, in fact, equal to the radius of the first Bohr orbit.

Since $(2l+1)$ corresponds to s in equation 9.21 and $(n+l)$ corresponds to k in that equation, equation 9.23 provides the relationship between n and l as

or
$$2l+1 \leqslant n+l$$
$$l \leqslant (n-1) \qquad (9.24)$$

Since the lowest value which l can take is zero, it follows from equation 9.24 that the lowest value of n is unity, therefore the permitted values of n are given by

$$n = 1, 2, 3, 4, \ldots \qquad (9.25)$$

THE ENERGY LEVELS

In the R, T and F equations, the total energy, E, of the system appeared in the R equation, equation 9.14. The quantum number n which appears in the associated Laguerre polynomials which are solutions of the R equation is related to the total energy. A full consideration of the R equation shows that the associated Laguerre polynomials are solutions provided that

$$n^2 = -\frac{2\pi^2 Z^2 m e^4}{(4\pi\varepsilon_0)^2 h^2 E}$$

from which

$$\boxed{E = -\frac{2\pi^2 Z^2 m e^4}{(4\pi\varepsilon_0)^2 n^2 h^2}} \qquad (9.26)$$

The energy of the hydrogen atom is thus quantized and depends on the quantum number n which is known as the *principal* quantum number. The two other quantum numbers, l and m, are known as the *azimuthal* and *magnetic* quantum numbers respectively. The possible values of all three quantum numbers may be summarised here by

$$n = 1, 2, 3, 4, \ldots \qquad (9.27)$$
$$l = 0, 1, 2, 3, \ldots, (n-1) \qquad (9.28)$$
$$m = 0, \pm 1, \pm 2, \pm 3, \ldots, \pm l \qquad (9.29)$$

The specification of a particular state of the atom requires the specification of the values of all three quantum numbers. The energy of the atom, however, is given by equation 9.26 and depends only on

128 THE HYDROGEN ATOM

the value of n. States with different values of l and m but with the same value of n are thus degenerate.

For a given value of n, equation 9.28 shows that there are n different possible values of l and equation 9.29 shows that there are $(2l+1)$ different possible values of m. The total number of states with the same energy is thus given by

$$\sum_{l=0}^{n-1}(2l+1) = 1+3+5+\ldots+(2n-1) = \frac{(1+2n-1)n}{2} = n^2 \quad (9.30)$$

There is thus n^2-fold degeneracy for the hydrogen atom.

It is important to appreciate that this degeneracy results from the fact that the potential energy of the electron is given by the Coulombic interaction of the single electron with the nucleus as specified in equation 9.1. In multi-electron atoms where there is interaction between the electrons in addition to electron–nucleus interaction, the energy of the atom will depend upon both n and l. In this case different values of l will lead to different energies. The magnetic quantum number m can, however, still have $(2l+1)$ different values for a given value of l, so that even under these conditions there will still exist $(2l+1)$ fold degeneracy. It is only in the presence of a magnetic field that this remaining degeneracy is removed and states with different values of m have different energies.

It may be worth pointing out that the effect of an electric field is a little different. In a magnetic field the energy of the atom depends on the value of m, but in an electric field the energy depends on the modulus of m so that the states $m = +1$ and $m = -1$, for example, will still be degenerate.

THE ANGULAR MOMENTUM OF THE ELECTRON

In the previous section it was seen that the principal quantum number controlled the energy of the isolated hydrogen atom and that the azimuthal quantum number made some contribution to the determination of the energy in multi-electronic atoms. The azimuthal quantum number is further associated with the angular momentum of the electron as also is the magnetic quantum number.

Before considering the angular momentum of a quantum mechanical system such as the hydrogen atom it will be useful to consider classical angular momentum. The angular momentum of a particle about a point is defined as the linear momentum multiplied by the perpendicular distance of the point from the line in which the particle is travelling. Suppose a particle of mass m travels in a circle of radius r in the xy plane, the centre of the circle being the origin of the

THE HYDROGEN ATOM 129

co-ordinate system. This situation is represented in *Figure 9.1*. If the linear velocity of the particle is v its linear momentum will be mv. Its instantaneous direction of travel is tangential to the circle and the perpendicular distance between the direction of travel and the origin is equal to the radius of the circle, r. The magnitude of the angular momentum of the particle is thus mvr. The anticlockwise direction about the origin is taken conventionally as positive so that the velocity of the particle illustrated in *Figure 9.1* will be a positive quantity and hence the angular momentum will be positive.

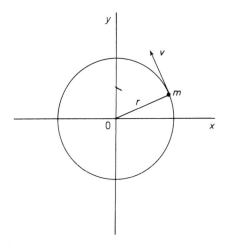

Figure 9.1   Angular momentum

The angular momentum about the origin may be represented by a vector, perpendicular to the plane of motion, through the origin. The right-hand screw rule applies to this representation so that for the situation illustrated in *Figure 9.1* the angular momentum vector would project upwards from the plane of the diagram along the z-axis. If the motion of the particle were reversed the angular momentum vector would project downwards representing a negative quantity. This concept is illustrated in *Figure 9.2*.

The angular momentum of the particle in *Figures 9.1* and *9.2* may be denoted by the symbol L_z which may be called the z component of the angular momentum. In this particular example the z component is, in fact, equal to the total angular momentum because the particle was supposed to move only in the xy plane.

For motion in the xy plane which is not necessarily circular, the z component of angular momentum may be expressed in terms of the

130 THE HYDROGEN ATOM

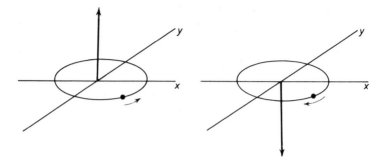

Figure 9.2   Vector representation of angular momentum

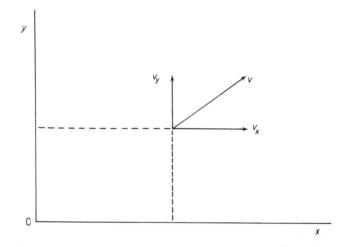

Figure 9.3   Angular momentum in terms of linear momentum

linear momenta of the particle parallel to the x- and y-axes. This point may be understood by reference to *Figure 9.3*.

The velocity, v, of a particle moving in the xy plane may be resolved into a component, v_y, parallel to the y-axis and a component, v_x, parallel to the x-axis. The linear momenta in these two directions will have the magnitudes mv_y and mv_x which may be denoted p_y and p_x respectively. The angular momentum about the origin due to p_y is obtained by multiplying p_y by the perpendicular distance from the origin to the line of the v_y velocity component and it may be seen from *Figure 9.3* that this distance is equal to the x co-ordinate of the particle. As v_y is in an anticlockwise direction about the origin

THE HYDROGEN ATOM

the resultant angular momentum is positive and is equal to xp_y. The angular momentum about the origin due to p_x is obtained in a similar way and as v_x is in a clockwise direction about the origin the resultant angular momentum is negative and is equal to $-yp_x$. The total angular momentum is equal to the sum of the two angular momenta computed above and hence

$$L_z = xp_y - yp_x \tag{9.31}$$

If the particle moved in three dimensions rather than in the xy plane only there would be components of angular momentum along the x- and y-axes also and these would be given by

$$L_x = yp_z - zp_y \tag{9.32}$$
$$L_y = zp_x - xp_z \tag{9.33}$$

and the total angular momentum, L, is related to the x, y and z components by

$$L^2 = L_x^2 + L_y^2 + L_z^2 \tag{9.34}$$

To determine the angular momentum of a quantum mechanical system the wave function may be operated on by the appropriate angular momentum operator. It should be remembered that a quantum mechanical operator is constructed by writing down the classical expression for the observable of interest in terms of co-ordinates and momenta and then replacing all momentum terms by the momentum operator $(h/2\pi i)(\partial/\partial q)$. Considering the z component of angular momentum, the classical expression in terms of co-ordinates and momenta is given by equation 9.31. To construct the z component angular momentum operator, denoted $\mathscr{L}_z$, the terms p_y and p_x in equation 9.31 are replaced by the appropriate form of the momentum operator when

$$\mathscr{L}_z = \frac{h}{2\pi i}\left(x\frac{\partial}{\partial y} - y\frac{\partial}{\partial x}\right) \tag{9.35}$$

The operators for the y and x components of angular momentum, $\mathscr{L}_y$ and $\mathscr{L}_x$ respectively, are similarly given by

$$\mathscr{L}_y = \frac{h}{2\pi i}\left(z\frac{\partial}{\partial x} - x\frac{\partial}{\partial z}\right) \tag{9.36}$$

$$\mathscr{L}_x = \frac{h}{2\pi i}\left(y\frac{\partial}{\partial z} - z\frac{\partial}{\partial y}\right) \tag{9.37}$$

Further, the operator for the square of the total angular momentum, $\mathscr{L}^2$, may be expressed in terms of the component operators,

$$\mathscr{L}^2 = \mathscr{L}_x^2 + \mathscr{L}_y^2 + \mathscr{L}_z^2 \tag{9.38}$$

132 THE HYDROGEN ATOM

To operate on the hydrogen atom wave function, the operators must be converted to spherical polar co-ordinates and for $\mathscr{L}_z$ and $\mathscr{L}^2$ the results are

$$\mathscr{L}_z = \frac{h}{2\pi i} \frac{\partial}{\partial \varphi} \tag{9.39}$$

$$\mathscr{L}^2 = -\frac{h^2}{4\pi^2} \left[\frac{1}{\sin\theta} \frac{\partial}{\partial \theta} \left(\sin\theta \frac{\partial}{\partial \theta} \right) + \frac{1}{\sin^2\theta} \frac{\partial^2}{\partial \varphi^2} \right] \tag{9.40}$$

Equation 9.39 was the operator used to calculate the momentum of the particle on a ring in Chapter 7 because the particle moved only in the xy plane.

Remembering that ψ is given by equation 9.5 as the product of the R, T and F equations

$$\psi = RTF$$

the operation of $\mathscr{L}^2$ on ψ may be represented

$$\mathscr{L}^2 \psi = -\frac{h^2}{4\pi^2} \left[\frac{1}{\sin\theta} \frac{\partial}{\partial \theta} \left(\sin\theta \frac{\partial}{\partial \theta} \right) + \frac{1}{\sin^2\theta} \frac{\partial^2}{\partial \varphi^2} \right] RTF$$

or

$$\mathscr{L}^2 \psi = -\frac{h^2}{4\pi^2} \left[\frac{RF}{\sin\theta} \frac{d}{d\theta} \left(\sin\theta \frac{dT}{d\theta} \right) + \frac{RT}{\sin^2\theta} \frac{d^2 F}{d\varphi^2} \right] \tag{9.41}$$

From the T equation, equation 9.13, and equation 9.17, the first term in the bracket in equation 9.41 is given by

$$\frac{RF}{\sin\theta} \frac{d}{d\theta} \left(\sin\theta \frac{dT}{d\theta} \right) = -RF \left(l(l+1) - \frac{m^2}{\sin^2\theta} \right) T \tag{9.42}$$

From the F equation, equation 9.12, the second term in the bracket of equation 9.41, is given by

$$\frac{RT}{\sin^2\theta} \frac{d^2 F}{d\varphi^2} = -\frac{RTFm^2}{\sin^2\theta} \tag{9.43}$$

Substituting from equations 9.42 and 9.43 into equation 9.41

$$\mathscr{L}^2 \psi = -\frac{h^2}{4\pi^2} \left[RTF \left(-l(l+1) + \frac{m^2}{\sin^2\theta} \right) - \frac{RTFm^2}{\sin^2\theta} \right]$$

$$= \frac{h^2}{4\pi^2} l(l+1) RTF$$

or

$$\mathscr{L}^2 \psi = \frac{h^2}{4\pi^2} l(l+1) \psi$$

The result of the operation has given a real number multiplied by the original wave function so that the real number is equal to the square of the total angular momentum. Hence

$$L^2 = \frac{h^2}{4\pi^2} l(l+1)$$

or

$$L = \sqrt{[l(l+1)]} \frac{h}{2\pi} \qquad (9.44)$$

from which it is seen that the azimuthal quantum number determines the total angular momentum in multiples of $h/2\pi$.

The operation of $\mathscr{L}_z$ on ψ may be represented by

$$\mathscr{L}_z \psi = \frac{h}{2\pi i} \frac{\partial}{\partial \varphi}(RTF)$$

or

$$\mathscr{L}_z \psi = \frac{h}{2\pi i} \cdot RT \frac{dF}{d\varphi} \qquad (9.45)$$

The function F is given by equation 9.15 as

$$F = \frac{1}{\sqrt{(2\pi)}} e^{im\varphi}$$

so that equation 9.45 becomes

$$\mathscr{L}_z \psi = \frac{h}{2\pi i} \cdot RT \frac{1}{\sqrt{(2\pi)}} \frac{d}{d\varphi}(e^{im\varphi})$$

$$= \frac{h}{2\pi i} \cdot RT \frac{im}{\sqrt{(2\pi)}} e^{im\varphi}$$

$$= \frac{mh}{2\pi} RTF$$

or

$$\mathscr{L}_z \psi = \frac{mh}{2\pi} \psi$$

Once again the operation has generated a real number multiplied by the original wave function so that

$$L_z = m \frac{h}{2\pi} \qquad (9.46)$$

from which it is seen that the magnetic quantum number determines the z component of the angular momentum in integral multiples of $h/2\pi$.

If $\mathscr{L}_x$ and $\mathscr{L}_y$ operate on the wave function, the result is not a real number multiplied by the original wave function so that $\mathscr{L}_x$ and $\mathscr{L}_y$ are not sharp quantities.

For a particular value of l, the maximum value of m is given by $m = l$ so that the maximum value of L_z is $lh/2\pi$. This is less than $\sqrt{[l(l+1)]}(h/2\pi)$ so that the z component of angular momentum must always be less than the total angular momentum. This is a consequence of the Uncertainty Principle. If L_z were equal to the total value, L, then both L_x and L_y would have to be zero. In this case both the magnitude of the angular momentum and the direction of the axis about which the motion was occurring would be precisely known. Further, if L_z were equal to L the electron would be confined to motion in the xy plane so that its orbit would be planar. As L_z must be less than L, planar orbits are not allowed in the quantum mechanical description of the hydrogen atom.

Another way of illustrating this fact is to consider that the linear momentum in the z direction, p_z, of a particle restricted to the xy plane must be zero. Additionally, the value of the z co-ordinate of the particle must also be zero. The uncertainty in the z co-ordinate, Δz, and the uncertainty in the linear momentum in the z direction, Δp_z, are thus both zero and hence $\Delta z \cdot \Delta p_z = 0$ which is a violation of the Uncertainty Principle.

From equations 9.44 and 9.46 it may be understood that not only is the total angular momentum limited to the values determined by the former equation, but the direction which the vector is allowed to take is restricted by the consideration that the z component must be an integral multiple of $h/2\pi$. This directional limitation is sometimes known as *space quantization*. This may be illustrated, for example, by reference to the situation where $l = 1$. For this example

$$L = \sqrt{2}\frac{h}{2\pi}$$

and

$$L_z = \frac{h}{2\pi}, \quad 0, \quad -\frac{h}{2\pi}$$

since

$$m = +1, 0, -1 \quad \text{for} \quad l = 1$$

The length of the total angular momentum vector is thus equal to $\sqrt{2}(h/2\pi)$. For $L_z = 0$, the L vector must lie somewhere in the xy plane. For $L_z = h/2\pi$ the L vector must lie in a conical surface and point upwards (in the positive z direction) and for $L_z = -h/2\pi$, the L vector must lie in a conical surface and point downwards in the

THE HYDROGEN ATOM   135

negative z direction. This situation is illustrated in *Figure 9.4(a)* with a three-dimensional representation, and *Figure 9.4(b)* where the representation is drawn in two dimensions to show how the lengths of the L_z and L vectors determine the angle of the cone in which the L vector lies. The L vector must be free to move in the conical surface because if its direction were fixed the implication would be that the electron moved in a planar orbit.

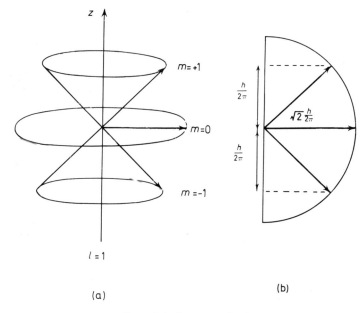

Figure 9.4   Space quantization

It should be noticed in passing that the choice of the z-axis as the axis of quantization is quite arbitrary and stems from the arbitrary way in which spherical polar co-ordinates are related to cartesian co-ordinates.

THE HYDROGEN WAVE FUNCTIONS

The total wave function $\psi(r, \theta, \varphi)$ is the product of the R, T and F functions as indicated by equation 9.4. The total wave function will depend upon all three quantum numbers, the R function depends upon n and l, the T function depends upon l and m and the F function

136 THE HYDROGEN ATOM

depends only on m. This statement is frequently summarised by

$$\psi_{n,l,m}(r, \theta, \varphi) = R_{n,l}(r) \cdot T_{l,m}(\theta) \cdot F_m(\varphi) \tag{9.47}$$

The eigenfunctions of the hydrogen atom are called orbitals and are named according to the values of n and l, the numerical value of n being followed by a letter which indicates the value of l according to

$$\begin{array}{cccccc} l = 0 & 1 & 2 & 3 & 4 & \ldots \\ s & p & d & f & g & \ldots \end{array}$$

When it is necessary to specify the value of m this may be done by adding a subscript to the orbital notation. For example, the orbital where $n = 2$, $l = 1$, $m = -1$, may be denoted $2p_{-1}$. Additionally, the wave function may be specified by giving the values for n and l in brackets. Thus the wave function for a 1s orbital is written $\psi(1, 0)$ and for a 3d orbital $\psi(3, 2)$.

The significance of the hydrogen orbitals is often considered by taking separately the *radial function*, R, and the *angular function*, TF. Frequently the angular function is represented by the single symbol Y, so that

$$T_{l,m}(\theta) \cdot F_m(\varphi) = Y_{l,m}(\theta, \varphi)$$

and

$$\psi_{n,l,m}(r, \theta, \varphi) = R_{n,l}(r) \cdot Y_{l,m}(\theta, \varphi) \tag{9.48}$$

THE RADIAL FUNCTION

It must be stressed again that the radial functions themselves have no physical significance. The quantity RR^* can, however, be used to calculate the probability of the electron being a certain distance, in any direction, from the nucleus. Since the radial functions do not contain complex numbers, RR^* is the same as R^2 and the quantity of interest is $R^2 \, dV$ which is the probability of finding the electron in the volume element dV in any direction from the nucleus. That is to say, it is the probability of finding the electron within a spherical shell of volume dV with the nucleus at the centre of the shell.

The volume of a spherical shell of radius r and thickness dr is $4\pi r^2 \, dr$ so that the probability of finding the electron between r and $(r+dr)$ from the nucleus is $4\pi r^2 R^2 \, dr$. This probability is usually represented by plotting $4\pi r^2 R^2$ against r which gives rise to the familiar radial probability distributions.

THE HYDROGEN ATOM   137

THE ANGULAR FUNCTION

The angular part of the wave function is important in respect of the formation of chemical bonds which depends on the suitable overlap of the orbitals of the individual atoms forming the bond. The overlap involves the *total* wave function, ψ, but whether or not suitable overlap occurs will depend on the shape of the orbitals which is governed by the angular functions which are sometimes known as spherical harmonics. For s orbitals where $l = 0$ the angular function is $Y_{0,0}$ and is given by

$$Y_{0,0} = \frac{1}{2\sqrt{\pi}} \qquad (9.49)$$

This function is spherically symmetrical.

For p orbitals where $l = 0$ and $m = +1, 0, -1$ there will be three angular functions corresponding to the three values of m. These are

$$Y_{1,0} = \frac{\sqrt{3}}{2\sqrt{\pi}} \cos\theta \qquad (9.50)$$

$$Y_{1,+1} = \frac{\sqrt{3}}{2\sqrt{(2\pi)}} \sin\theta \cdot e^{i\varphi} \qquad (9.51)$$

$$Y_{1,-1} = \frac{\sqrt{3}}{2\sqrt{(2\pi)}} \sin\theta \cdot e^{-i\varphi} \qquad (9.52)$$

As $Y_{1,0}$ is a real function it can be plotted and represented by a diagram and this can be achieved in more than one way. One method, for example, would be to plot the cosine function with the cartesian y-axis representing $Y_{1,0}$ and the cartesian x-axis representing θ. This method would simply generate the normal cosine curve between $\theta = 0$ and $\theta = 2\pi$, the amplitude of the curve being $(\sqrt{3})/2(\sqrt{\pi})$.

The more usual method of representing the angular functions is by a polar plot, as shown in *Figure 9.5*. For a particular value of θ a line of length $[(\sqrt{3})/2(\sqrt{\pi})] \cos\theta$ is drawn from the origin so that it makes an angle θ with the z-axis. As $Y_{1,0}$ is independent of φ, this line may take any orientation with respect to the x-axis and thus generates a circle, the plane of the circle being parallel to the xy plane. Varying θ from 0 to 2π will thus generate the surfaces of two spheres with their centres lying on the z-axis, as shown in *Figure 9.5*. It is important to realise that this plot has no physical significance, it is merely a method of representing the angular function $Y_{1,0}$ by a diagram.

Although these polar plots of the angular functions are those most commonly encountered, they can be misleading if they are not correctly interpreted. The difficulties may be overcome by a simpler method of representation.

From the point of view of the formation of chemical bonds the important consideration is the extent of overlap of the two atomic orbitals of the atoms which are bonded together. It is the directional properties of the orbitals which can be used to give a description of

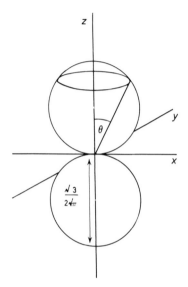

Figure 9.5 Polar diagram of the $Y_{1,0}$ function

directed valence and it is therefore only necessary to know in which direction an orbital has its maximum magnitude. If the $Y_{1,0}$ function is considered, equation 9.50 shows that it has a maximum value when $\cos \theta$ is a maximum, i.e., when $\theta = 0$. Reference to *Figure 7.1*, which illustrates the relationship between cartesian and polar co-ordinates, shows that the direction corresponding to $\theta = 0$ lies along the z-axis. The directions of maximum magnitude of the $Y_{1,0}$ function may thus be represented by arrows drawn along the z-axis, as shown in *Figure 9.6*. It is important to note that the actual values of the function are positive above the xy plane and negative below it. These signs arise from the values of $\cos \theta$ in these regions and this antisymmetry is important in determining the values of the overlapped orbitals in a chemical bond.

The functions $Y_{1,+1}$ and $Y_{1,-1}$ cannot be represented by a diagram because they are complex and contain an imaginary part. The imaginary part arises from the *F* equation which was first discussed in Chapter 7 for the particle on a ring. It was pointed out at the end of that chapter that it is a property of degenerate wave functions that any linear combination of the members of a degenerate set is also an eigenfunction of the same energy as the original wave

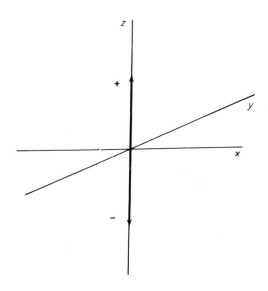

Figure 9.6 Directional properties of the $Y_{1,0}$ function

functions. As $Y_{1,+1}$ and $Y_{1,-1}$ are degenerate, a linear combination of them will be an eigenfunction, and this linear combination is formed in exactly the same way as equations 7.21 and 7.23 were generated in Chapter 7. Thus,

$$\frac{1}{\sqrt{2}}(Y_{1,+1} + Y_{1,-1}) = \frac{1}{\sqrt{2}} \frac{\sqrt{3}}{2\sqrt{(2\pi)}} \sin\theta(e^{i\varphi} + e^{-i\varphi})$$

$$= \frac{\sqrt{3}}{2\sqrt{\pi}} \sin\theta \cos\varphi \qquad (9.53)$$

and

$$\frac{1}{i\sqrt{2}}(Y_{1,+1} - Y_{1,-1}) = \frac{1}{i\sqrt{2}} \frac{\sqrt{3}}{2\sqrt{(2\pi)}} \sin\theta(e^{i\varphi} - e^{-i\varphi})$$

$$= \frac{\sqrt{3}}{2\sqrt{\pi}} \sin\theta \sin\varphi \qquad (9.54)$$

140 THE HYDROGEN ATOM

The direction of the maximum magnitudes of the functions in equations 9.53 and 9.54 can be represented diagrammatically by the same method as was applied to the function $Y_{1,0}$. Thus, equation 9.53 has a maximum value when $\sin\theta \cos\varphi$ is a maximum, i.e., when $\theta = \pi/2$ and $\varphi = 0$. This direction lies along the x-axis. Similarly, the maximum value of equation 9.54 occurs when $\sin\theta \sin\varphi$ is a maximum, i.e., when $\theta = \pi/2$ and $\varphi = \pi/2$ and this direction lies along the y-axis.

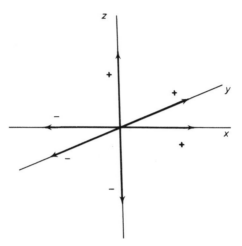

Figure 9.7 Directional properties of p orbitals

The complex functions $Y_{1,+1}$ and $Y_{1,-1}$ have thus been combined to produce two real functions in much the same way as an imaginary displacement was converted to a real displacement in the discussion of simple harmonic motion in Chapter 2.

The maximum values of the functions represented by equations 9.50, 9.53 and 9.54 may thus be represented on one diagram as shown in *Figure 9.7* from which it may be seen that the directions of the maximum values of the angular functions are directed at mutual right angles.

This spatial orientation can be used to give a description of bond angles in various molecules.

The fact that the maximum values of the angular functions lie in the x, y and z directions results from the fact that the angular dependence of the functions given in equations 9.50, 9.53 and 9.54 are the same as the angular dependence of the z, x and y co-ordinates respectively. This may be seen from the relationship between cartesian and spherical polar co-ordinates which were given in Chapter 7 by

$$x = r \sin \theta \cos \varphi$$
$$y = r \sin \theta \sin \varphi \quad (9.55)$$
$$z = r \cos \theta$$

The p orbitals of which equations 9.50, 9.53 and 9.54 give the angular dependence, are thus labelled p_z, p_x and p_y respectively. Remembering that the total wave function contains the radial function, the orbitals for p_x, p_y and p_z may be expressed by

$$p_x = \frac{1}{\sqrt{2}}(Y_{1,+1} + Y_{1,-1})R_{2,1} \quad (9.56)$$

$$p_y = \frac{1}{i\sqrt{2}}(Y_{1,+1} - Y_{1,-1})R_{2,1} \quad (9.57)$$

$$p_z = Y_{1,0} \cdot R_{2,1} \quad (9.58)$$

Notice that whilst $m = 0$ for the p_z orbital, the p_x and p_y orbitals each have some $m = +1$ and $m = -1$ character.

Further to this point it is of interest to consider the angular probability functions of the p orbitals. This function for the $Y_{1,0}$ angular function is given by $Y_{1,0} \cdot Y_{1,0}^*$. As $Y_{1,0}$ is a real function, $Y_{1,0}^*$ is the same as $Y_{1,0}$ and the angular probability function is obtained from equation 9.50 as

$$Y_{1,0}^2 = \frac{3}{4\pi} \cos^2 \theta \quad (9.59)$$

The functions $Y_{1,+1}$ and $Y_{1,-1}$ are complex because they contain an imaginary part. The probability functions are real, however, and are obtained from equations 9.51 and 9.52 as

$$Y_{1,+1} \cdot Y_{1,+1}^* = \frac{\sqrt{3}}{2\sqrt{(2\pi)}} \sin \theta \cdot e^{i\varphi} \left(\frac{\sqrt{3}}{2\sqrt{(2\pi)}} \sin \theta \cdot e^{-i\varphi} \right)$$

or

$$Y_{1,+1} \cdot Y_{1,+1}^* = \frac{3}{8\pi} \sin^2 \theta \quad (9.60)$$

Similarly,

$$Y_{1,-1} \cdot Y_{1,-1}^* = \frac{3}{8\pi} \sin^2 \theta \quad (9.61)$$

It should be noted that the probability functions given by equations 9.59, 9.60 and 9.61 are all independent of φ and are thus symmetrical about the z-axis. Furthermore, the functions corresponding to $m = +1$ and $m = -1$ are identical, the former corresponding to anticlockwise motion of the electron about the z-axis and the latter corresponding to clockwise motion.

As

$$p_z = Y_{1,0} \cdot R_{2,1}$$

the angular probability functions for p_z or $Y_{1,0}$ are the same. In the case of p_x and p_y which are mixtures of $Y_{1,+1}$ and $Y_{1,-1}$, however, the situation is rather different and may at first sight present a difficulty. Representing the angular probability functions of p_x and p_y as Y_x^2 and Y_y^2 respectively, equations 9.53 and 9.54 give

$$Y_x^2 = \frac{3}{4\pi} \sin^2 \theta \cos^2 \varphi \qquad (9.62)$$

and

$$Y_y^2 = \frac{3}{4\pi} \sin^2 \theta \sin^2 \varphi \qquad (9.63)$$

and these probability functions obviously differ from those of equations 9.60 and 9.61. When it is remembered that p_x and p_y are mixtures of $Y_{1,+1}$ and $Y_{1,-1}$, however, it seems more reasonable to compare the angular probability functions

$$(Y_x^2 + Y_y^2) \quad \text{and} \quad (Y_{1,+1} \cdot Y_{1,+1}^* + Y_{1,-1} \cdot Y_{1,-1}^*)$$

These totals may be derived from equations 9.62 and 9.63 and from equations 9.60 and 9.61 respectively.

$$(Y_x^2 + Y_y^2) = \frac{3}{4\pi} \sin^2 \theta \cos^2 \varphi + \frac{3}{4\pi} \sin^2 \theta \sin^2 \varphi$$

$$= \frac{3}{4\pi} \sin^2 \theta (\cos^2 \varphi + \sin^2 \varphi)$$

or

$$(Y_x^2 + Y_y^2) = \frac{3}{4\pi} \sin^2 \theta \qquad (9.64)$$

Also

$$(Y_{1,+1} \cdot Y_{1,+1}^* + Y_{1,-1} \cdot Y_{1,-1}^*) = \frac{3}{8\pi} \sin^2 \theta + \frac{3}{8\pi} \sin^2 \theta$$

$$= \frac{3}{4\pi} \sin^2 \theta \qquad (9.65)$$

It may thus be seen that the angular probability functions derived from $Y_{1,+1}$ together with $Y_{1,-1}$ and from p_x with p_y are the same. Moreover, they are both symmetrical about the z-axis.

In this respect it is interesting to note that in compounds where double π bonds are formed the distribution of electrons in the orbitals is cylindrically symmetrical about the bond axis. In nitrogen for example, where a σ bond is formed by the axial overlap of a p_z orbital from each atom the p_x and p_y orbitals are at right angles to each other and also at right angles to the σ bond. They thus form two π bonds and the electron distribution in the nitrogen molecule is cylindrically symmetrical about the nitrogen–nitrogen axis.

The points which have been discussed above in relation to p orbitals apply also to d orbitals and with the $2d$ orbitals, for example, the angular function for $m = 0$ will be real.

$$Y_{2,0} = N'(3\cos^2\theta - 1) \tag{9.66}$$

where N' is the normalisation constant. As $\cos^2\theta$ is the square of the angular dependence of the z co-ordinate (equation 9.55) this orbital is labelled d_{z^2}. Thus

$$d_{z^2} = Y_{2,0}$$

The orbitals for $m = \pm 1, \pm 2$ are given by

$$\begin{aligned} Y_{2,\pm 1} &= N\sin 2\theta \cdot e^{\pm i\varphi} \\ Y_{2,\pm 2} &= N\sin^2\theta \cdot e^{\pm 2i\varphi} \end{aligned} \tag{9.67}$$

where N is a normalisation constant. These complex orbitals are converted to real functions by taking linear combinations and the cartesian subscripts are derived from the same considerations as applied to the p orbitals.

$$\begin{aligned} d_{xz} &= \frac{1}{\sqrt{2}}(Y_{2,+1} + Y_{2,-1}) \\ d_{yz} &= \frac{1}{i\sqrt{2}}(Y_{2,+1} - Y_{2,-1}) \\ d_{x^2-y^2} &= \frac{1}{\sqrt{2}}(Y_{2,+2} + Y_{2,-2}) \\ d_{xy} &= \frac{1}{i\sqrt{2}}(Y_{2,+2} - Y_{2,-2}) \end{aligned} \tag{9.68}$$

It should be noted that for a set of degenerate orbitals, *any* linear combination is an acceptable function. It is necessary, however, that the members of any set of degenerate orbitals be orthogonal to one another. Thus $Y_{2,0}$, $Y_{2,\pm 1}$, and $Y_{2,\pm 2}$ are orthogonal to one another and constitute an acceptable set of orbitals. Also the orbitals given in equations 9.68 are all orthogonal to one another and constitute a further acceptable set. It is, however, not acceptable to mix members from each set since d_{xz} and $Y_{2,+1}$, for example, are not orthogonal to one another.

ELECTRON SPIN

In Chapter 1 it was pointed out that a fourth quantum number was required to specify the state of an electron in an atom and this necessity arose from the observation that many spectroscopic lines were doublets, consisting of two lines very close together. Goudsmit and Uhlenbeck postulated that an electron had an intrinsic angular momentum which contributes to the total angular momentum. This intrinsic angular momentum is often called *spin* by analogy with the classical situation where a body of finite size spinning about its own axis would have an intrinsic angular momentum.

The spin of the electron is treated in a manner analogous to the orbital angular momentum. A quantum number, s, is associated with the total spin angular momentum, S, and a quantum number, m_s, is associated with the z component of the spin S_z. The quantum numbers s and m_s are the spin analogues of l and m. Thus for orbital angular momentum there exist the relationships

$$L = \sqrt{[l(l+1)]}\frac{h}{2\pi}$$

$$L_z = m\frac{h}{2\pi}$$

and, by analogy, for spin angular momentum

$$\boxed{S = \sqrt{[s(s+1)]}\frac{h}{2\pi}} \qquad (9.69)$$

$$\boxed{S_z = m_s\frac{h}{2\pi}} \qquad (9.70)$$

To account for the experimental observations the value of the quantum number s must be restricted to the single value of $\frac{1}{2}$ and the quantum number m_s can only take the values of $+\frac{1}{2}$ and $-\frac{1}{2}$.

As mentioned in Chapter 1, this *ad hoc* assumption is unnecessary in Dirac's treatment of quantum mechanics where the fourth quantum number m_s emerges naturally and is used to specify the symmetry or antisymmetry of the wave function. Dirac's approach involves using the complete time-dependent Schrödinger equation and considering the motion of the electron from the point of view of relativity.

APPENDIX 1

COMPLEX NUMBERS

A complex number is one which has a real component and an imaginary component. It may be written

$$a + ib \tag{A1.1}$$

where a and b are real numbers and $i = (-1)^{1/2}$.

Alternatively, a complex number may be expressed as an exponential such as e^{ix}, where x is a real number. Expanding this exponential

$$e^{ix} = 1 + ix - \frac{x^2}{2!} - \frac{ix^3}{3!} + \frac{x^4}{4!} + \frac{ix^5}{5!} \ldots$$

or

$$e^{ix} = \left(1 - \frac{x^2}{2!} + \frac{x^4}{4!} \ldots\right) + i\left(x - \frac{x^3}{3!} + \frac{x^5}{5!} \ldots\right) \tag{A1.2}$$

The series in the brackets in equation A1.2 are the expansions of $\cos x$ and $\sin x$, hence

$$\boxed{e^{ix} = \cos x + i \sin x} \tag{A1.3}$$

Similarly,

$$\boxed{e^{-ix} = \cos x - i \sin x} \tag{A1.4}$$

Adding equations A1.3 and A1.4 gives

$$2 \cos x = e^{ix} + e^{-ix}$$

or

$$\cos x = \tfrac{1}{2}(e^{ix} + e^{-ix}) \tag{A1.5}$$

Subtracting equation A1.4 from A1.3

$$2i \sin x = e^{ix} - e^{-ix}$$

or

$$\sin x = \frac{1}{2i}(e^{ix} - e^{-ix}) \tag{A1.6}$$

Having defined the sine and cosine functions in terms of complex exponentials, it is worth noting at this point that the hyperbolic functions $\sinh x$ and $\cosh x$ are defined by similar relations with i omitted. Thus

$$\cosh x = \tfrac{1}{2}(e^{x} + e^{-x}) \tag{A1.7}$$

and

$$\sinh x = \tfrac{1}{2}(e^{x} - e^{-x}) \tag{A1.8}$$

Further,

$$\frac{\sinh x}{\cosh x} = \tanh x \tag{A1.9}$$

and

$$\frac{\cosh x}{\sinh x} = \coth x \tag{A1.10}$$

Now consider the complex number $re^{i\alpha}$ where both r and α are real. From equation A1.3

$$re^{i\alpha} = r(\cos \alpha + i \sin \alpha) \tag{A1.11}$$

Comparing this statement with equation A1.1 it may be understood that the complex number $(a+ib)$ may be expressed as $re^{i\alpha}$ provided certain relationships hold between a and b, and r and α. Thus putting

$$a + ib = re^{i\alpha}$$

gives

$$a + ib = r(\cos \alpha + i \sin \alpha)$$

so that

$$a = r \cos \alpha$$

APPENDIX 1 147

and (A1.12)
$$b = r\sin\alpha$$

This conversion from the form $(a+ib)$ to the form $re^{i\alpha}$ provides a good method for multiplying complex numbers. To determine the value of $(a+ib)(c+id)$ for example, put

$$a+ib = r_1 e^{i\alpha_1}$$

and

$$c+id = r_2 e^{i\alpha_2}$$

Then

$$(a+ib)(c+id) = r_1 e^{i\alpha_1} \cdot r_2 e^{i\alpha_2} = r_1 r_2 e^{i(\alpha_1+\alpha_2)}$$

THE CONJUGATES OF COMPLEX NUMBERS

Consider the complex number $re^{i\alpha}$ and the complex number $re^{-i\alpha}$. The second is called the conjugate of the first. Alternatively, if a complex number is written $(a+ib)$, its conjugate is $(a-ib)$. In general, the conjugate of a complex number is obtained by replacing i by $-i$.

Conjugate complex numbers have two important properties;

(i) the sum of a complex number and its conjugate is real, e.g.,

$$(a+ib)+(a-ib) = 2a \qquad (A1.13)$$

or

$$re^{i\alpha} + re^{-i\alpha} = r(e^{i\alpha} + e^{-i\alpha})$$

Hence, from equation A1.5

$$re^{i\alpha} + re^{-i\alpha} = 2r\cos\alpha \qquad (A1.14)$$

(ii) the product of a complex number and its conjugate is real, e.g.,

$$re^{i\alpha} \cdot re^{-i\alpha} = r^2 e^{i(\alpha-\alpha)} = r^2$$

From equation A1.12

$$a = r\cos\alpha$$
$$b = r\sin\alpha$$

Hence

$$r^2 = a^2 + b^2 \qquad (A1.15)$$

and

$$\boxed{(a+ib)(a-ib) = a^2 + b^2} \qquad (A1.16)$$

APPENDIX 2

THE TUNNEL EFFECT

Equation 4.47 is derived by eliminating D_1, C_2 and D_2 from equations 4.43, 4.44, 4.45 and 4.46 which are stated as

$$C_1 + D_1 = C_2 + D_2 \tag{4.43}$$

$$k_1(C_1 - D_1) = k_2(C_2 - D_2) \tag{4.44}$$

$$C_2 e^{ik_2 a} + D_2 e^{-ik_2 a} = C_3 e^{ik_3 a} \tag{4.45}$$

$$k_2(C_2 e^{ik_2 a} - D_2 e^{-ik_2 a}) = k_3 C_3 e^{ik_3 a} \tag{4.46}$$

From equation 4.43

$$C_1 = C_2 + D_2 - D_1 \tag{A2.1}$$

and from equation 4.44

$$C_1 - D_1 = \frac{k_2}{k_1}(C_2 - D_2)$$

or

$$D_1 = C_1 - \frac{k_2}{k_1}(C_2 - D_2) \tag{A2.2}$$

Substituting this expression for D_1 into equation A2.1

$$C_1 = C_2 + D_2 - C_1 + \frac{k_2}{k_1}(C_2 - D_2)$$

whence

$$2C_1 = C_2\left(1 + \frac{k_2}{k_1}\right) + D_2\left(1 - \frac{k_2}{k_1}\right)$$

or
$$C_1 = \tfrac{1}{2}C_2\left(1+\frac{k_2}{k_1}\right)+\tfrac{1}{2}D_2\left(1-\frac{k_2}{k_1}\right) \tag{A2.3}$$

From equation 4.45
$$C_2 = (C_3 e^{ik_3 a} - D_2 e^{-ik_2 a}) \cdot e^{-ik_2 a} \tag{A2.4}$$

From equation 4.46
$$C_2 e^{ik_2 a} - D_2 e^{-ik_2 a} = \frac{k_3}{k_2} C_3 e^{ik_3 a}$$

whence
$$C_2 e^{ik_2 a} = \frac{k_3}{k_2} C_3 e^{ik_3 a} + D_2 e^{-ik_2 a}$$

or
$$C_2 = \left(\frac{k_3}{k_2} C_3 e^{ik_3 a} + D_2 e^{-ik_2 a}\right) \cdot e^{-ik_2 a} \tag{A2.5}$$

Equating the two expressions for C_2 given by equations A2.4 and A2.5

$$e^{-ik_2 a}(C_3 e^{ik_3 a} - D_2 e^{-ik_2 a}) = e^{-ik_2 a}\left(\frac{k_3}{k_2} C_3 e^{ik_3 a} + D_2 e^{-ik_2 a}\right)$$

which leads to
$$C_3 e^{ik_3 a}\left(1 - \frac{k_3}{k_2}\right) = 2 D_2 e^{-ik_2 a}$$

or
$$D_2 = \tfrac{1}{2} C_3 \left(1 - \frac{k_3}{k_2}\right) \cdot e^{ik_3 a} \cdot e^{ik_2 a} \tag{A2.6}$$

Substituting for D_2 from equation A2.6 into equation A2.4

$$C_2 = \left[C_3 e^{ik_3 a} - \tfrac{1}{2} C_3 \left(1 - \frac{k_3}{k_2}\right) \cdot e^{ik_3 a} \cdot e^{ik_2 a} \cdot e^{-ik_2 a}\right] e^{-ik_2 a}$$

$$C_2 = C_3 e^{ik_3 a} \cdot e^{-ik_2 a}\left[1 - \tfrac{1}{2}\left(1 - \frac{k_3}{k_2}\right)\right]$$

or
$$C_2 = \tfrac{1}{2}\left(1 + \frac{k_3}{k_2}\right) C_3 e^{ik_3 a} \cdot e^{-ik_2 a} \tag{A2.7}$$

APPENDIX 2

Substituting from equations A2.6 and A2.7 for D_2 and C_2 into equation A2.3

$$C_1 = \tfrac{1}{2}\left(1+\frac{k_2}{k_1}\right)\left[\tfrac{1}{2}\left(1+\frac{k_3}{k_2}\right)C_3\,e^{ik_3 a}\cdot e^{-ik_2 a}\right] +$$

$$+ \tfrac{1}{2}\left(1-\frac{k_2}{k_1}\right)\left[\tfrac{1}{2}\left(1-\frac{k_3}{k_2}\right)C_3\,e^{ik_3 a}\cdot e^{ik_2 a}\right]$$

$$C_1 = \tfrac{1}{4}C_3\,e^{ik_3 a}\left[\left(1+\frac{k_2}{k_1}\right)\left(1+\frac{k_3}{k_2}\right)e^{-ik_2 a} + \left(1-\frac{k_2}{k_1}\right)\left(1-\frac{k_3}{k_2}\right)e^{ik_2 a}\right]$$

$$C_1 = \tfrac{1}{4}C_3\,e^{ik_3 a}\left[\left(1+\frac{k_3}{k_2}+\frac{k_2}{k_1}+\frac{k_3}{k_1}\right)e^{-ik_2 a} + \left(1-\frac{k_3}{k_2}-\frac{k_2}{k_1}+\frac{k_3}{k_1}\right)e^{ik_2 a}\right]$$

$$C_1 = \tfrac{1}{4}C_3\,e^{ik_3 a}\left[\left(1+\frac{k_3}{k_1}\right)e^{-ik_2 a} + \left(1+\frac{k_3}{k_1}\right)e^{ik_2 a} +\right.$$

$$\left.+\left(\frac{k_3}{k_2}+\frac{k_2}{k_1}\right)e^{-ik_2 a} - \left(\frac{k_3}{k_2}+\frac{k_2}{k_1}\right)e^{ik_2 a}\right]$$

$$C_1 = \tfrac{1}{4}C_3\,e^{ik_3 a}\left[\left(1+\frac{k_3}{k_1}\right)(e^{ik_2 a}+e^{-ik_2 a}) - \left(\frac{k_3}{k_2}+\frac{k_2}{k_1}\right)(e^{ik_2 a}-e^{-ik_2 a})\right]$$

$$C_1 = \tfrac{1}{4}C_3\,e^{ik_3 a}\left[2\left(1+\frac{k_3}{k_1}\right)\cosh(ik_2 a) - 2\left(\frac{k_3}{k_2}+\frac{k_2}{k_1}\right)\sinh(ik_2 a)\right]$$

or finally,

$$C_1 = \tfrac{1}{2}C_3\,e^{ik_3 a}\left[\left(1+\frac{k_3}{k_1}\right)\cosh(ik_2 a) - \left(\frac{k_3}{k_2}+\frac{k_2}{k_1}\right)\sinh(ik_2 a)\right] \quad (4.47)$$

APPENDIX 3

TRIGONOMETRIC RELATIONSHIPS

The trigonometric relationships used in this book are mostly derived from or directly employ the relationships

$$\sin(x+y) = \sin x \cos y + \cos x \sin y \quad \text{(A3.1)}$$
$$\cos(x+y) = \cos x \cos y - \sin x \sin y \quad \text{(A3.2)}$$
$$\cos(x-y) = \cos x \cos y + \sin x \sin y \quad \text{(A3.3)}$$

Putting $x = y$ in equation A3.2 yields

$$\cos 2x = \cos^2 x - \sin^2 x \quad \text{(A3.4)}$$

and remembering that

$$\cos^2 x + \sin^2 x = 1 \quad \text{(A3.5)}$$
$$\cos^2 x = 1 - \sin^2 x$$

Substituting in equation A3.4

$$\cos 2x = 1 - 2\sin^2 x$$

whence

$$\boxed{\sin^2 x = \tfrac{1}{2}(1 - \cos 2x)} \quad \text{(A3.6)}$$

Alternatively, from equation A3.5

$$\sin^2 x = 1 - \cos^2 x$$

and substituting in equation A3.4

$$\cos 2x = 2\cos^2 x - 1$$

whence

$$\boxed{\cos^2 x = \tfrac{1}{2}(\cos 2x - 1)} \quad \text{(A3.7)}$$

Subtracting equation A3.2 from equation A3.3 gives

$$\cos(x-y) - \cos(x+y) = 2\sin x \sin y$$

or, alternatively,

$$\boxed{\sin x \sin y = \tfrac{1}{2}\cos(x-y) - \tfrac{1}{2}\cos(x+y)} \qquad (A3.8)$$

Putting $x = y$ in equation A3.1 gives

$$\sin 2x = \sin x \cos x + \cos x \sin x$$

or

$$\boxed{\sin x \cos x = \tfrac{1}{2}\sin 2x} \qquad (A3.9)$$

Remember also that

$$\tan x = \frac{\sin x}{\cos x} = \frac{1}{\cot x} \qquad (A3.10)$$

and further, from Appendix 1 that

$$\sin x = \frac{1}{2i}(e^{ix} - e^{-ix}) \qquad (A3.11)$$

$$\cos x = \tfrac{1}{2}(e^{ix} + e^{-ix}) \qquad (A3.12)$$

APPENDIX 4

DIFFERENTIALS AND INTEGRALS

The differentiation and integration used in this book employs the following relationships:

$$\frac{d}{dx}(r \sin ax) = ar \cos ax \quad (A4.1)$$

$$\frac{d}{dx}(r \cos ax) = -ar \sin ax \quad (A4.2)$$

$$\int (r \sin ax) \, dx = -\frac{r}{a} \cos ax \quad (A4.3)$$

$$\int (r \cos ax) \, dx = \frac{r}{a} \sin ax \quad (A4.4)$$

$$\frac{d}{dx}(r e^{ax}) = ar e^{ax} \quad (A4.5)$$

$$\int (r e^{ax}) \, dx = \frac{r}{a} e^{ax} \quad (A4.6)$$

Notice that in all the above relationships, simpler cases may be obtained by putting a or r or both equal to unity.

DIFFERENTIATION OF A PRODUCT

If u and v are each functions of the variable x, then

$$\frac{d}{dx}(uv) = u\frac{dv}{dx} + v\frac{du}{dx} \quad (A4.7)$$

Only one other form will be required and this is

$$\frac{d}{dx}(r e^{ax^2})$$

154 APPENDIX 4

Let $y = r e^{ax^2}$ when the required differential will be dy/dx. Put

$$z = ax^2$$

when

$$\frac{dz}{dx} = 2ax$$

and

$$y = r e^z$$

From equation A4.5

$$\frac{dy}{dz} = r e^z$$

Further,

$$\frac{dy}{dx} = \frac{dy}{dz} \cdot \frac{dz}{dx}$$

Hence

$$\frac{dy}{dx} = r e^z \cdot 2ax$$

Substituting for z, the required differential is given by

$$\frac{d}{dx}(r e^{ax^2}) = 2axr e^{ax^2} \qquad (A4.8)$$

FURTHER READING

The following list contains some suggestions for further reading which will provide an expansion and extension of the material covered in this book. The list is by no means comprehensive, but the books quoted will give further references to more specialised topics.

LINNETT, J. W. (1960). *Wave Mechanics and Valency*, Methuen, London.
MURRELL, J. N., KETTLE, S. F. A., and TEDDER, J. M. (1965). *Valence Theory*, Wiley, London.
BARRETT, J. (1970). *Introduction to Atomic and Molecular Structure*, Wiley, London.
KARPLUS, M., and PORTER, R. N. (1970). *Atoms and Molecules*, W. A. Benjamin, New York.
BARROW, G. M. (1962). *Introduction to Molecular Spectroscopy*, McGraw-Hill, New York.
HANNA, M. W. (1969). *Quantum Mechanics in Chemistry*, 2nd edn, W. A. Benjamin, New York.
KAUFMANN, E. D. (1966). *Advanced Concepts in Physical Chemistry*, McGraw-Hill, New York.
BROWN, D. A. (1972). *Quantum Chemistry*, Penguin, Harmondsworth.
ATKINS, P. W. (1970). *Molecular Quantum Mechanics*, Clarendon Press, Oxford.

INDEX

α-particles
  scattering of, 1
Absorption bands, 73
  electronic spectra, 71
Angular function, 136–143
  polar plots, 137
Angular momentum, 101, 109, 129, 130
  electron, 3, 128–135
  orbital, 144
  total, 130, 133, 134
Angular momentum component, 134
Angular momentum operator, 109, 131
Angular momentum vector, 129
Angular probability functions, 142
Anharmonic oscillator, 100
Anharmonicity constant, 100
Antibonding, 87
Antisymmetric functions, 84–88, 138
Arbitrary constants, 15, 21, 25, 26
Associated Laguerre polynomial, 126, 127
Associated Legendre polynomial, 116, 118, 119
Asymptotic solutions, 95
Average values, 41, 45
Azimuthal quantum number, 127, 133

Barriers
  potential energy, 54–70
    penetration of, 65
    single, 57–65
Black body radiator, 2
Bohr, 3, 4, 6
Bonding, 87, 89, 137
Boundary conditions, 21, 29
Box
  one-dimensional, 28–31, 72, 78

  three-dimensional, 49–53
de Broglie, 5, 6
  equation, 23

Chemical bonds, 137, 138
Complex exponential solutions, 17, 25, 54, 55, 103
Complex numbers, 145–147
  conjugate, 147
Compton, 3
Conjugated polyenes, 74
Conjugated systems, 71–74
Construction of operators, 41
Correspondence principle, 31, 98
Coulomb's law, 80

Davisson, 5
Degeneracy, 53, 110, 120, 128, 139, 143
Diatomic molecule
  electronic energy of, 80–90
  vibrational energy of, 74–80
Differential wave equation, 19, 22
Differentials, 153, 154
Diffraction of electrons, 5
Dirac, 5, 7, 144
Directional implications of wave functions, 54–56, 58
Directional properties of p-orbitals, 140
d-orbitals, 143
Double well model, 83

Eigenfunctions, 24–26, 29, 30
  normalised, 33
Eigenvalue equation, 44
Eigenvalues, 24–26, 30
Electron angular momentum, 128–135
Electron diffraction, 5

158 INDEX

Electron spin, 144
Electronic energy of a diatomic molecule, 80–90
Electronic spectra absorption bands, 71
Electronic spectrum, 72
Energy
  kinetic, 10
  potential, 10, 25, 80, 91, 122
  total, 10
  zero point, 30, 98
Energy and amplitude, 10, 11, 26
Energy levels
  antisymmetric states, 90
  hydrogen atom, 127, 128
  isolated states, 90
  linear harmonic oscillator, 98, 99
  symmetric states, 90

Force constant, 11, 91
Frequency, 5

Geiger, 1
Germer, 5
Goudsmit, 6, 144

Hamilton, 6
Hamiltonian operator, 42, 43
Harmonic oscillator, 91–100
Heisenberg, 4, 26
Hermite polynomial, 95–97
Hermitian operators, 41, 47, 48
Hydrogen
  atom, 122–144
    energy levels, 127, 128
  molecule ion, 80, 82
  wave functions, 135, 136

Integrals, 153, 154
Interpretation of ψ, 26, 27

Kinetic energy, 10

Laguerre polynomial, 126
  associated, 126, 127
Laplacian operator, 24, 101, 102, 113, 122
Legendre polynomial, 116
  associated, 116, 118, 119
Lennard, 1
Linear combinations, 12, 14, 21, 110, 139, 143
Linear harmonic oscillator, 91–100
Linear operators, 36, 41

Magnetic quantum number, 6, 127, 133
Marsden, 1
Maxwell, 2
Millikan, 3
Moment of inertia, 113
Momentum
  angular, 101, 109, 129, 134
  of particle in a box, 55
  operator, 44, 45, 56, 131
Morse curve, 74, 79, 91, 92, 99

Node, 30
Non-rigid rotator, 121
Normalisation, 32, 33, 41, 108
Normalised eigenfunctions, 33

One-dimensional box, 28–31, 72, 78
One-dimensional models, 71–90
Operators, 36, 37
  angular momentum, 109, 131
  Hamiltonian, 42, 43
  Hermitian, 41, 47, 48
  Laplacian, 24, 101, 102, 113, 122
  linear, 36, 41
  momentum, 44, 45, 56, 131
  quantum mechanical, 36–38, 41
Orbital angular momentum, 144
Orbital overlap, 137
Orthogonal pairs, 35
Orthogonality, 33–36, 143
Orthonormal functions, 35

Particle
  density, 26
  in a one-dimensional box, 28–31, 71
  in a three-dimensional box, 49–53
  on a ring, 101–110
Periodicity
  spatial, 18
  temporal, 9, 10, 18
Photoelectric effect, 3, 5
Photons, 3, 5
Planck, 2, 3, 6
Polar co-ordinates, 101, 102
Polar diagrams, 138
Polar plots of angular functions, 137
Polyenes
  conjugated, 74
Polynomials
  associated Laguerre, 126, 127
  associated Legendre, 116, 118, 119
  Hermite, 95–97
  Laguerre, 126
  Legendre, 116

INDEX

p-orbitals, 137, 141, 143
   directional properties of, 140
   probability functions of, 141
Postulates of quantum mechanics, 38–49
Potential energy, 10, 25, 80, 91, 122
   barriers, 54–70
   penetration of, 65
   single, 57–65
Principal quantum number, 127
Probability, 26
   distributions, 30, 31, 141, 142
   radial, 136
Progressive wave equation, 21
Progressive waves, 17–20

Quantum mechanical operators, 36–38, 41
Quantum number, 3, 4, 6, 30, 119
   azimuthal, 127, 133
   magnetic, 6, 127, 133
   principal, 127
   rotational, 119, 120
   spin, 144
   vibrational, 98
Quantum theory, 3, 4

Radial function, 136
Radial probability distribution, 136
Recursion formula, 97
Reid, 5, 26
Rigid rotator, 111–121
Rotational quantum number, 119, 120
Rotational spectra, 120
Rupp, 6
Rutherford, 1, 2

Schrödinger, 5, 6, 21
   time dependent equation, 144
   time independent equation, 24–40
   wave equation, 22–24
Separation of variables, 50, 113, 115, 124
Sharp quantities, 43, 56
Simple harmonic motion, 8–10
   general differential equation of, 11, 12
Single potential barriers, 57–65
Sommerfeld, 4
Spatial periodicity, 18
Spherical harmonics, 137
Spherical polar co-ordinates, 101, 102
Spin, 144
   quantum number, 144
Standing waves, 19–22

States of system, 44–52
   degenerate, 53, 110, 120, 128, 139
Stern, 6
Symmetric functions, 84, 85, 87, 88

Temporal periodicity, 9, 10, 18
Thomson, G. P., 5, 26
Thomson, J. J., 1
Three-dimensional box, 49–53
Time dependent Schrödinger equation, 144
Time independent Schrödinger equation, 24, 40
Total angular momentum, 130, 133
Total angular momentum vector, 134
Total energy, 10
Transmission probability, 67
Transparency factor, 68
Trigonometric relationships, 151, 152
Trigonometric solutions, 16, 25, 55, 56, 103
Tunnel effect, 65–70, 148–150

Uhlenbeck, 6, 144
Uncertainty principle, 26, 47
Unsharp quantities, 41

Vector
   angular momentum, 129
   total angular momentum, 134
Vibrational energy
   quantization of, 77
Vibrational energy of a diatomic molecule, 74–80
Vibrational quantum number, 98

Wave
   equations, 22
      Schrödinger, 22–24
      simple harmonic motion, 8–10
   functions, 84
      antisymmetric, 84–88
      directional implications of, 54–56, 58
      symmetric, 84, 85, 87, 88
Wavelength, 5
Waves
   progressive, 17–21
   standing, 19–22
Well-behaved functions, 24

Zeemann, 4, 53
Zero point energy, 30, 98